Lecture Notes in Electrical Engineering

Volume 368

About this Series

"Lecture Notes in Electrical Engineering (LNEE)" is a book series which reports the latest research and developments in Electrical Engineering, namely:

- Communication, Networks, and Information Theory
- Computer Engineering
- Signal, Image, Speech and Information Processing
- Circuits and Systems
- Bioengineering

LNEE publishes authored monographs and contributed volumes which present cutting edge research information as well as new perspectives on classical fields, while maintaining Springer's high standards of academic excellence. Also considered for publication are lecture materials, proceedings, and other related materials of exceptionally high quality and interest. The subject matter should be original and timely, reporting the latest research and developments in all areas of electrical engineering.

The audience for the books in LNEE consists of advanced level students, researchers, and industry professionals working at the forefront of their fields. Much like Springer's other Lecture Notes series, LNEE will be distributed through Springer's print and electronic publishing channels.

More information about this series at http://www.springer.com/series/7818

James J. (Jong Hyuk) Park
Gangman Yi · Young-Sik Jeong
Hong Shen
Editors

Advances in Parallel and Distributed Computing and Ubiquitous Services

UCAWSN & PDCAT 2015

 Springer

Editors

James J. (Jong Hyuk) Park
Department of Computer Science and
  Engineering
Seoul University of Science and Technology
Seoul
Korea, Republic of (South Korea)

Gangman Yi
Department of Computer Science and
  Engineering
Gangneung-Wonju National University
Wonju
Korea, Republic of (South Korea)

Young-Sik Jeong
Department of Multimedia Engineering
Dongguk University
Seoul
Korea, Republic of (South Korea)

Hong Shen
School of Computer Science
University of Adelaide
Adelaide, SA
Australia

ISSN 1876-1100   ISSN 1876-1119 (electronic)
Lecture Notes in Electrical Engineering
ISBN 978-981-10-0067-6   ISBN 978-981-10-0068-3   (eBook)
DOI 10.1007/978-981-10-0068-3

Library of Congress Control Number: 2015954979

Springer Singapore Heidelberg New York Dordrecht London

Printed on acid-free paper

Springer Science+Business Media Singapore Pte Ltd. is part of Springer Science+Business Media
(www.springer.com)

Message from the UCAWSN 2015 General Chairs

The 4th International Conference on Ubiquitous Computing Application and Wireless Sensor Network (UCAWSN-15) is an event of the series of international scientific conferences. This conference takes place in Jeju, Korea, July 8–10, 2015. The UCAWSN-15 will be the most comprehensive conference focused on the various aspects of Ubiquitous Computing Application and Wireless Sensor Network (UCAWSN). The UCAWSN-15 will provide an opportunity for academic and industry professionals to discuss the latest issues and progress in the area of UCAWSN. In addition, the conference will publish high quality papers which are closely related to the various theories and practical applications. Furthermore, we expect that the conference and its publications will be a trigger for further related research and technology improvements in this important subject.

The papers included in the proceedings cover two tracks: Track 1—Ubiquitous Computing and Track 2—Wireless Sensor Network. Accepted and presented papers highlight new trends and challenges of Ubiquitous Computing Application and Wireless Sensor Network. The presenters showed how new research could lead to novel and innovative applications. We hope you will find these results useful and inspiring for your future research.

We would like to express our sincere thanks to the Program Chairs: Gangman Yi (Gangneung-Wonju National University, Korea), Yuh-Shyan Chen (National Taipei University, Taiwan), Xiaohong Peng (Aston University, UK), Ali Abedi (Maine University, US), Neil Y. Yen (the University of Aizu, Japan), Jen Juan Li (North Dakota State University, USA), Hongxue (Harris) Wang (Athabasca University, Canada), Lijun Zhu (ISTIC, China), and all Program Committee members and all the additional reviewers for their valuable efforts in the review process, which helped us to guarantee the highest quality of the selected papers for the conference. Our special thanks go to the invited speaker who kindly accepted our invitations, and helped to meet the objectives of the conference: Prof. Hong Shen (The University of Adelaide, Australia) and Prof. Doo-soon Park

(SoonChunHyang University, Korea). We cordially thank all the authors for their valuable contributions and the other participants of this conference. The conference would not have been possible without their support. Thanks are also due to the many experts who contributed to making the event a success.

UCAWSN 2015 General Chairs

C.S. Raghavendra, University of Southern California, USA
Young-SikJeong, Dongguk University, Korea
Jason C. Hung, Overseas Chinese University, Taiwan
Hanmin Jung, KISTI, Korea
Yang Xiao, University of Alabama, USA

Message from the UCAWSN 2015 Program Chairs

Welcome to the 4th International Conference on Ubiquitous Computing Application and Wireless Sensor Network (UCAWSN-15), which is held in Jeju on July 8–10, 2015. UCAWSN 2015 is the most comprehensive conference focused on the various aspects of information technology. UCAWSN 2015 provides an opportunity for academic and industry professionals to discuss the latest issues and progress in the area of UCAWSN such as Ubiquitous and context-aware computing, context-awareness reasoning and representation, locations awareness services, architectures, protocols and algorithms of WSN, energy, management and control of WSN, etc. In addition, the conference will publish high-quality papers which are closely related to the various theories and practical applications in UCAWSN. Furthermore, we expect that the conference and its publications will be a trigger for further related research and technology improvements in this important subject.

For UCAWSN 2015, we received many paper submissions, after a rigorous peer review process, we accepted the articles with high quality for the UCAWSN 2015 proceedings. All submitted papers have undergone blind reviews by at least three reviewers from the technical program committee, which consists of leading researchers around the globe. Without their hard work, achieving such a high-quality proceeding would not have been possible. We take this opportunity to thank them for their great support and cooperation. We would like to sincerely thank the following general chairs who helped to meet the objectives of the conference: Prof. Young-SikJeong (Dongguk University, Korea) and Hwa-Young Jeong (Kyung Hee University, Korea). Finally, we would like to thank all of you for your participation in our conference, and also thank all the authors, reviewers, and organizing committee members. Thank you and enjoy the conference!

UCAWSN 2015 Program Chairs

Gangman Yi, Gangneung-Wonju National University, Korea
Yuh-Shyan Chen, National Taipei University, Taiwan
Xiaohong Peng, Aston University, UK

Ali Abedi, Maine University, USA
Neil Y. Yen, the University of Aizu, Japan
Jen Juan Li, North Dakota State University, USA
Hongxue (Harris) Wang, Athabasca University, Canada
Lijun Zhu, ISTIC, China

Organization

Honorary Chair
Doo-soon Park, SoonChunHyang University, Korea

General Chairs
Hwa-Young Jeong, Kyung Hee University, Korea
Han-Chieh Chao, National Ilan University, Taiwan
Yi Pan, Georgia State University, USA
QunJin, Waseda University, Japan

General Vice-Chairs
Young-SikJeong, Dongguk University, Korea
Jason C. Hung, Overseas Chinese University, Taiwan
Hanmin Jung, KISTI, Korea
Yang Xiao, University of Alabama, USA

Program Chairs
Gangman Yi, Gangneung-Wonju National University, Korea
Yuh-Shyan Chen, National Taipei University, Taiwan
Xiaohong Peng, Aston University, UK
Ali Abedi, Maine University, USA
Neil Y. Yen, the University of Aizu, Japan
Jen Juan Li, North Dakota State University, USA
Hongxue (Harris) Wang, Athabasca University, Canada
Lijun Zhu, ISTIC, China

International Advisory Board
SheraliZeadally, University of the District of Columbia, USA
Naveen Chilamkurti, La Trobe University, Australia
Luis Javier Garcia Villalba, Universidad Complutense de Madrid (UCM), Spain
Mohammad S. Obaidat, Monmouth University, USA
Jianhua Ma, Hosei University, Japan
James J. Park, SeoulTech, Korea

Laurence T. Yang, St. Francis Xavier University, Canada
Hai Jin, HUST, China
Hamid R. Arabnia, The University of Georgia, USA
WeijiaJia, City U. of Hong Kong, Hong Kong
Albert Zomaya, University of Sydney, Australia
Bin Hu, Lanzhou University, China
Doo-soon Park, SoonChunHyang University, Korea

Publicity Chairs
Deok-Gyu Lee, Seowon University, Korea
Byung-Gyu Kim, Sunmoon University, Korea
Weiwei Fang, Beijing Jiaotong University, China
Cain Evans, Birmingham City University, UK
Chun-Cheng Lin, National Chiao Tung University, Taiwan
Antonio Coronato, ICAR, Italy
Rung-Shiang Cheng, Kunshan University, Taiwan
Haixia Zhang, Shandong University, China
Rafael Falcon, Larus Technologies, Canada
Xu Shao, Institute for Infocomm Research, Singapore
Bong-Hwa Hong, Kyunghee Cyber University, Korea
Hak Hyun Choi, Seoul Women's University, Korea

Local Arrangement Chairs
Namje Park, Jeju National University, Korea
Cheonshik Kim, Sejong University, Korea
Min Choi, Chungbuk National University, Korea
Aziz Nasridinov, Dongguk University, Korea

Message From the PDCAT 2015 General Chairs

The 16th International Conference on Parallel and Distributed Computing, Applications and Technologies (PDCAT) is a major forum for scientists, engineers, and practitioners throughout the world to present their latest research, results, ideas, developments, and applications in all areas of parallel and distributed computing. Beginning in Hong Kong in 2000, PDCAT-15 will be held in Jeju, Korea after 15 years of successful journey through various countries/regions including Taiwan, Japan, China, Singapore, Australia, New Zealand, and Korea across Asia-Oceania. We are inviting new and unpublished papers.

The conference papers included in the proceedings cover the following topics: PDCAT of Networking and Architectures, Software Systems and Technologies, Algorithms and Applications, and Security and Privacy. Accepted and presented papers highlight new trends and challenges of Parallel and Distributed Computing, Applications and Technologies. We hope readers will find these results useful and inspiring for their future research.

We would like to express our sincere thanks to Steering Chair: Hong Shen (University of Adelaide, Australia). Our special thanks go to the Program Chairs: Joon-Min Gil (Catholic University of Daegu, Korea) and Ching-Hsien (Robert) Hsu (Chung Hua University, Taiwan), and all Program Committee members and all reviewers for their valuable efforts in the review process that helped us to guarantee the highest quality of the selected papers for the conference.

PDCAT 2015 General Chairs

James Park, SeoulTech, Korea
Hamid R. Arabnia, University of Georgia, USA
Han-Chieh Chao, National Ilan University, Taiwan
Albert Y. Zomaya, University of Sydney, Australia

Organization Committee

Honorary Chair
Doo-Soon Park, KIPS President/SoonChunHyang University, Korea

Steering Chair
Hong Shen, University of Adelaide, Australia

General Chairs
James Park, SeoulTech, Korea
Hamid R. Arabnia, University of Georgia, USA
Han-Chieh Chao, National Ilan University, Taiwan
Albert Y. Zomaya, University of Sydney, Australia

General Vice-Chairs
Young-SikJeong, Dongguk University, Korea
Michael Jeong, Kyung Hee University, Korea
Hanmin Jung, KISTI, Korea

Program Chairs
Joon-Min Gil, Catholic University of Daegu, Korea
Weisong Shi, Wayne State University, USA
Ching-Hsien (Robert) Hsu, Chung Hua University,Taiwan

Workshop Chairs
Gangman Yi, Gangneung-Wonju National University, Korea
Francis Lau, University of Hong Kong, Hong Kong
Jason C. Hung, Overseas Chinese University, Taiwan
KaLok Man, Xi'an Jiaotong-Liverpool University, China
Neil Y. Yen, The University of Aizu, Japan

Publicity Chairs
Deqing Zou, HUST, China
Deok-Gyu Lee, Seowon University, Korea

Julio Sahuquillo, Universidad Politecnica de Valencia, Spain
Akihiro Fujiwara, Kyushu Institute of Technology, Japan
Bo-Chao Cheng, National Chung-Cheng University, Taiwan

Local Arrangement Chairs
Namje Park, Jeju National University, Korea
Hwamin Lee, Soonchunhyang University, Korea

Contents

Rhymes+: A Software Shared Virtual Memory System
with Three Way Coherence Protocols on the Intel Single-Chip
Cloud Computer 1
C.-H. Dominic Hung

Review and Comparison of Mobile Payment Protocol 11
Pensri Pukkasenung and Roongroj Chokngamwong

POFOX: Towards Controlling the Protocol Oblivious
Forwarding Network 21
Xiaodong Tan, Shan Zou, Haoran Guo and Ye Tian

An Experimental Study on Social Regularization
with User Interest Similarity 31
Zhiqi Zhang and Hong Shen

Representing Higher Dimensional Arrays into Generalized
Two-Dimensional Array: G2A 39
K.M. Azharul Hasan and Md Abu Hanif Shaikh

A Portable and Platform Independent File System for Large Scale
Peer-to-Peer Systems and Distributed Applications 47
Andreas Barbian, Stefan Nothaas, Timm J. Filler and Michael Schoettner

OCLS: A Simplified High-Level Abstraction Based Framework
for Heterogeneous Systems 57
Shusen Wu, Xiaoshe Dong, Heng Chen and Bochao Dang

Hierarchical Caching Management for Software Defined
Content Network Based on Node Value 67
Jing Liu, Lei Wang, Yuncan Zhang, Zhenfa Wang and Song Wang

**Interoperation of Distributed MCU Emulator/Simulator
for Operating Power Simulation of Large-Scale Internet
of Event-Driven Control Things**.   75
Sanghyun Lee, Bong Gu Kang, Tag Gon Kim, Jeonghun Cho
and Daejin Park

**The Greedy Approach to Group Students for Cooperative
Learning**.   83
Byoung Wook Kim, Sung Kyu Chun, Won Gyu Lee
and Jin Gon Shon

**Secure Concept of SCADA Communication for Offshore
Wind Energy**   91
Seunghwan Ju, Jaekyoung Lee, Joonyoung Park and Junshin Lee

**ASR Error Management Using RNN Based Syllable Prediction
for Spoken Dialog Applications**   99
Byeongchang Kim, Junhwi Choi and Gary Geunbae Lee

**A Protection Method of Mobile Sensitive Data and Applications
Over Escrow Service**. 107
Su-Wan Park, Deok Gyu Lee and Jeong Nyeo Kim

**GPU-Based Fast Refinements for High-Quality Color Volume
Rendering**. 117
Byeonghun Lee, Koojoo Kwon and Byeong-Seok Shin

**Beacon Distance Measurement Method in Indoor Ubiquitous
Computing Environment**. 125
Yunsick Sung, Jeonghoon Kwak, Young-Sik Jeong
and Jong Hyuk Park

**Indoor Location-Based Natural User Interface for Ubiquitous
Computing Environment**. 131
Jeonghoon Kwak and Yunsick Sung

**Flexible Multi-level Regression Model for Prediction of Pedestrian
Abnormal Behavior**. 137
Yu-Jin Jung and Yong-Ik Yoon

**Automatic Lighting Control Middleware System Controlled
by User's Emotion Based on EEG** 145
SoYoung Ahn, DongKyoo Shin, DongIl Shin and ChulGyun Park

Hand Recognition Method with Kinect. 153
DoYeob Lee, Dongkyoo Shin and Dongil Shin

**A Study on the Connectivity Patterns of Individuals
Within an Informal Communication Network**. 161
Somayeh Koohborfardhaghighi, Dae Bum Lee and Juntae Kim

Grid Connected Photovoltaic System Using Inverter. 167
HyunJong Kim, Moon-Taek Cho and Kab-Soo Kim

The Cluster Algorithm for Time-Varying Nonlinear System
with a Model Uncertainty 173
Jong-Suk Lee and Jong-Sup Lee

Integrated Plant Growth Measurement System Based on Intelligent
Circumstances Recognition 179
Moon-Taek Cho, Hae-Jong Joo and Euy-Soo Lee

A Study on the Big Data Business Model for the Entrepreneurial
Ecosystem of the Creative Economy. 185
Hyesun Kim, Mangyu Choi, Byunghoon Jeon and Hyoungro Kim

Implementation of Intelligent Decision-Based Smart
Group Scheduler. 191
Kyoung-Sup Kim, Yea-Bok Lee, Yi-Jun Min and Sang-Soo Kim

Implementation of MCA Rule Mapper for Cloud Computing
Environments 197
Kyoung-Sup Kim, Joong-il Woo, Jung-Eun Kim and Dong-Soo Park

A Simple Fatigue Condition Detection Method by using Heart
Rate Variability Analysis. 203
U.-Seok Choi, Kyoung-Ju Kim, Sang-Seo Lee, Kyoung-Sup Kim
and Juntae Kim

Insider Detection by Analyzing Process Behaviors of File Access. 209
Xiaobin Wang, Yongjun Wang, Qiang Liu, Yonglin Sun
and Peidai Xie

Analysis of the HOG Parameter Effect on the Performance
of Vision-Based Vehicle Detection by Support Vector Machine
Classifier 221
Kang Yi, Seok-Il Oh and Kyeong-Hoon Jung

A Fast Algorithm to Build New Users Similarity List
in Neighbourhood-Based Collaborative Filtering. 229
Zhigang Lu and Hong Shen

Rhymes+: A Software Shared Virtual Memory System with Three Way Coherence Protocols on the Intel Single-Chip Cloud Computer

C.-H. Dominic Hung

Abstract This research focuses on one prominent many-core prototype—the Intel's Single-chip Cloud Computer (SCC). We address the performance problem of shared virtual memory consistency for this cache in-coherent architecture. Aiming to keep data on-chip as much as possible to reduce memory accesses external to the chip, we propose two techniques to leverage the cache hierarchy to its full and make data reside in the on-chip scratchpad memory. First, targeted at the architectural specificity of the hardware, we redesigned the traditional software distributed shared memory (SDSM) to allow shared data to be treated transparently like private memory so that the cache hierarchy can be fully utilised without sacrificing memory consistency. Second, we propose a distance-aware page allocation scheme that samples access frequencies and selects the most frequently-recently used pages to be stored on the on-chip scratchpad memory.

Keywords Computer architecture · Many-core architecture · Shared virtual memory · Scratchpad · Distributed shared memory · Cache coherence

1 Introduction

With the advancements in integrated circuits technology, it is foreseeable that processors with more than 1,000 cores per die will appear in the near future. However, these many-core architectures have introduced many challenges at the memory system level, which are due to such special properties as complicated cache coherence and limited memory access speed.

The Intel's Single-chip Cloud Computer (SCC) is a prominent many-core prototype that does not provide hardware cache coherency. Instead, it relies on on-chip

C.-H. Dominic Hung (✉)
Department of Computer Science, The University of Hong Kong,
Pokfulam, Hong Kong
e-mail: chdhung@cs.hku.hk

© Springer Science+Business Media Singapore 2016   1
J.J.(Jong Hyuk) Park et al. (eds.), *Advances in Parallel and Distributed Computing and Ubiquitous Services*, Lecture Notes in Electrical Engineering 368,
DOI 10.1007/978-981-10-0068-3_1

programmable memory and manually handled cache coherency. Intel offered the Software Managed Coherence (SMC) layer as a software means for managing cache coherency. To achieve memory consistency, it accesses shared memory without the typical cache hierarchy for more efficient invalidation and flushing operations. We found that performance provided by such a coherence layer is sub-optimal because accesses of shared memory would all turn into data update messages within the network mesh. As cache locality could not be exploited to its full potential, the execution pipelines stall frequently for memory fetches from outside the chip.

We propose two methods that allow us to leverage the cache hierarchy to the full. The first method applies ordinary software distributed shared memory method, and the second method utilises the on-chip programmable memory to try to cache shared memory on the die area while maintaining consistency. We also mimic the Intel SMC, calling it SPM, to serve as the baseline protocol for comparison. All shared memory pages are allowed to swap between all three protocols dynamically in order to suit their shared memory access pattern.

Our experimental results show that our first technique, the ordinary SDSM outperforms the current SMC approach **by 5 times**. Moreover, in some cases, with the second technique that is based on scratchpad memory, our proposed system can improve the performance further **by an additional 1.57 times**. Our experiments also demonstrated that the SMC approach is not scalable due to congestion of the network mesh by coherence traffic generated while the two new approaches continued to scale well.

The main contribution of this research is the implementation of a cache coherence software library system for an architecture that comes with non-coherent cache hardware and relies only on software-defined cache. This new approach to cache hierarchy opens the door for smarter and faster inter-processor-core data sharing without the need of complicated cache coherence hardware.

2   Related Works

Numerous works have been done for the SCC platform to address shared memory issues. The first category of works targeted at message passing usage, such as RCCE [1]. The second class of works features shared virtual memory system, including Intel SMC [2] which is the baseline for comparison with our system.

More elaborated works include MetalSVM [3] which implemented an SVM system at the hypervisor level, the MESH [4] project that implemented an object-based SDSM system and SNU's CRF-based SVM [5] that implemented an SDSM system without much modern DSM optimisation techniques. The third type of works falls within language support for parallel programming. MultiMLton [6] and X10 on SCC [7] are projects that have implemented the functional language ML and PGAS language X10 for the Intel SCC.

3   Intel SCC

The Intel SCC features 24 tiles linked up by a 2-dimensional mesh, with each tile containing two x86 processor cores. Each processor core contains an integrated cache hierarchy and a globally accessible software-managed scratchpad. The scratchpad is on the same level as the L2$ in terms of design and speed if mapped to the layers of caches [8]. Cache-ability for L2$ and the scratchpad are mutually exclusive. This is because L1$ has been tailored to enable fast invalidation and eviction of a specific class of cache-lines by a single software-issued instruction. To be specific, the lines are tagged by an MPBT-bit which can be invalidated within a processor cycle by the instruction CL1INVMB which is specific to Intel SCC.

4   Three Way Coherence

The Rhymes+ library system supports three coherence protocols that are applied in a per-page per-core manner. Conversion between protocols can be performed on-the-fly depending on the expected reuse rate of a shared memory page. The Shared Physical Memory model targets at pages of lowest reuse rate, thus saving coherence operations overheads. A page can be configured manually to run under the DSM model when the expected reuse rate is high. The Shared Physical Memory-on-Chip model is applied automatically to pages of the highest reuse rate according to sampling data obtained by the system at runtime.

4.1   Shared Physical Memory

The Shared Physical Memory (SPM) model mimics the Intel SMC mode of operation to serve as the baseline coherence model for comparison. At acquire time, a core will first trigger the CL1INVMB instruction to invalidate any shared memory cache-lines in L1$. This assures that any subsequent read operations on a line within a shared memory page under the SPM protocol would fetch from the master copy that resides in the shared physical memory. Throughout a coherence scope, i.e., the interval between a lock and an unlock or that between two barriers, shared memory handled under the SPM model will only be cached by the L1$. In order to guarantee a line within a shared memory page under Rhymes+ is properly flushed from L1$ back to the physical memory at the point of release, the library system will make a write access to a special address to force the write-combining buffer to flush any outstanding writes.

4.2 Distributed Shared Memory

The Distributed Shared Memory (DSM) model is designed to target at shared memory pages of higher data re-use rate by leveraging the proximate cache hierarchy to the full, both L1$ and L2$, at the cost of coherence overheads. During the stage of acquire, the core will invalidate the working copy of a shared memory page if the page has already been cached locally and found modified elsewhere in the system.

Within the coherence scope, if an access is made to a page under the DSM operation model, a page fault is trapped. That leads to the duplication of the master copy of the page within the shared section of the system memory to the section private to the core. The duplication creates a working copy which memory accesses are made to, in lieu of the master copy. Rhymes+ incorporated the write vector technique such that modifications can be merged into the working copy in a fine-grained manner instead of removing the cached working copy as a whole if it already exists. The working copy will be configured as non-MPBT which makes it cache-able in the L2$ structure and can enjoy the cache effect that the L2$ brings. If the access is of the type write, one extra duplication of the master copy is required to generate an entrance snapshot of the page. That is used to contrast modifications at the end of a coherence scope in order to flush back to the master copy corresponding changes.

4.3 Shared Physical Memory-on-Chip

The Shared Physical Memory-on-Chip (SPM-OC) is designed for pages with even higher data re-use rates than those designated for DSM with the hope that the scratchpad layer can replace the L2$ in providing equivalent level of proximity. This reduces, but not obliterates, the overheads incurred from duplication and de-duplication under the DSM protocol by adopting an SPM-like model of operations.

The use of the SPM-OC model, unlike the manual switching from SPM to DSM, is automatic. Whether a page is to be swapped to run the SPM-OC protocol is decided by a heuristic that determines if a page running under the DSM model is the hottest among all other pages. Upon a fault on a page under the DSM protocol, which is eligible to be elevated to the SPM-OC protocol, the system will duplicate a working copy to reside in the scratchpad memory with consideration of the placement location so as to minimise the distance between the physical page slot containing the working copy and the cluster of cores that access the page often. All pages that have mapped to the same page running under the SPM protocol will be forced to un-map the page and re-map to the working copy in the scratchpad.

Meanwhile, those mapped the page under the DSM protocol will not be interrupted until release time. By then, their modifications will be flushed to the working copy in the scratchpad. Flushing alterations to this copy instead of the master copy can be benefited from the faster on-die memory. Subsequent faults on the same page for all cores will be directed to map the working copy in scratchpad regardless of the original designated protocol.

The current implementation allows elevation when the scratchpad has empty slots and does not allow spilling-out. De-elevation decisions are only made collectively at the end of a barrier interval with the aid of access frequency data sampled by the system. When a page is determined to be de-elevated, the working copy residing in the scratchpad will replace the master copy instance in physical memory; the detail mode of operation is same as SPM. In Fig. 1, we present the state diagram of the SPM-OC protocol with states of the SPM and DSM models that are allowed to transit to SPM-OC included, while other SPM and DSM states are hidden. The actual arrangement of the scratchpad for all practical purposes is to group page slots into associative sets for faster placement decision. In order to avoid imprecise judgement in elevating pages at cold start, a maturity period in terms of barrier count is required before actual allocation of pages into the scratchpad.

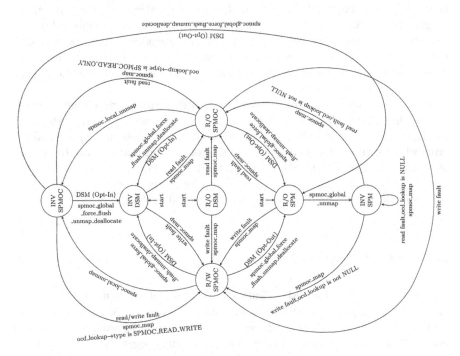

Fig. 1 Transition diagram for shared physical memory-on chip page

5 Performance Evaluations

Our benchmarking of the system made use of a modified matrix multiplication benchmark program from the UPC NAS parallel benchmark suite developed by the George Washington University. It calculates $C = AB^n$, with $n = 21$ and square matrices A and B of varying order m to demonstrate the scaling effect.

It should be noted that beyond the problem size of 64×64, the number of pages is no longer containable within the scratchpad of the Intel SCC. The DSM performance is obtained by setting the maturity period as infinite, making pages never be possible to elevate to SPM-OC. The discussion of the performance results obtained from the benchmark MM will be divided into three parts: the small problem size, the medium and large problem sizes; the performance results obtained showed different characteristics.

5.1 Small Problem Size

In small problem size trials (a) and (b), the actual shared memory footprint is 3–6 pages. It can be observed that with such a small problem size, the benchmark program has negative impact when being forced to run with more number of cores. One should be aware that, parallelising a problem brings overheads, for instance, barrier wait time or spin lock wait time that require computation to hide. The small problem size MM is clearly not one that can provide enough computation to hide such overheads (Fig. 2).

Furthermore, with the use of software DSM, further fixed overheads are incurred due to the duplication and de-duplication of shared memory pages. As these overheads cannot be absorbed by the computation execution, the baseline SPM protocol outperforms the DSM and SPM-OC protocols regardless. It can further be observed that SPM-OC in general outperforms the DSM protocol as after the scratchpad becomes mature, there will no longer be overheads induced by the DSM protocol when all shared memory pages are migrated to the scratchpad.

5.2 Medium Problem Size

The medium problem size MM (c) and (d) bears an actual shared memory footprint of 24 and 96 pages. It should be noted that the current associative scratchpad configuration can only provide 92 page slots, which means starting from trial of order 128, the scratchpad cannot keep all shared memory pages. That is to say, four pages will be left running under the DSM protocol when they are excluded from the scratchpad.

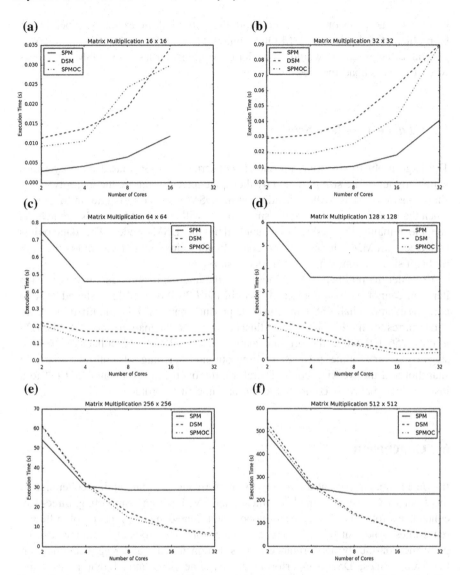

Fig. 2 Performance of matrix multiplication problem in all protocols. **a** Order 16, **b** Order 32, **c** Order 64, **d** Order 128, **e** Order 256, **f** Order 512

Starting from this group of trials, DSM and SPM-OC overtake the baseline SPM protocol. The larger pool of banks in L2$ avoided spill-outs and benefited the execution because intermittent spill-outs would stall the execution pipeline frequently when the small L1$ cannot contain all required shared memory. A cache miss will then mean fetching the corresponding data from the main memory. As the SCC allows only one outstanding write at the mesh interface unit, when there are heavy write accesses or cache contention, the stalling of the execution pipeline can be problematic. Again, the SPM-OC counterpart outperforms the DSM protocol in general as the execution would not suffer or would suffer less from DSM overheads when the scratchpad matures.

5.3 Large Problem Size

The large problem size trials (e) and (f) concern 384–1536 shared memory pages which are out of the size provided by the scratchpad. Therefore, the majority of the shared memory pages will be handled in the DSM way. To re-iterate, in SPM-OC, when the scratchpad can afford the space, pages will be migrated into the scratchpad and the remaining pages will be handled in the DSM model. The data points collected with MM running 2 and 4 cores do not follow the trend as in the medium problem size section, which worth discussion.

Consider the problem size of 256×256, the intermediate product matrix at each iteration cannot be held completely within the L2$. For workload shared by two cores as example, half of the intermediate product matrix will be modified by a core and requires the modifications to be flushed back into the main memory. As the L2$ is only 256 KB while the original matrix costs 512 KB, the collective flushing staged at barrier time requires spilling-out and fetching of both the entrance snapshot and the working copy. Therefore, the overheads dominated and led to a loss in performance as compared to the baseline SPM protocol.

6 Conclusion

In conclusion, targeted at improving the performance of maintaining coherence in the Intel SCC, we presented Rhymes+, an SVM system supporting three-way coherence protocols. The three protocols suit shared memory pages of different reuse rates. And that the changing of a page from one protocol to another can be performed on-the-fly. Performance results demonstrated that for applications with high reuse rates, DSM outperforms the baseline SPM model that mimics the Intel SMC, and that SPM-OC outperforms the DSM model while saving DSM overheads.

Acknowledgements This research is funded by the Research Grant Council of Hong Kong (HKU 7167/12E) and the Huawei Corporation. We thank Dr. Stefan Lankes and Mr. Pablo Reble of RWTH-Aachen University and Soft Tech Dev Beijing for their support in offering access to their Intel SCC.

References

1. Mattson TG, Riepen M, Lehnig T, Brett P, Haas W, Kennedy P, Howard J, Vangal S, Borkar N, Ruhl G, Dighe S (2010) The 48-core SCC processor: the programmer's view. In: Proceedings of the ACM/IEEE international conference for high performance computing, networking, storage and analysis, Nov 2010, pp 1–11
2. Zhou X, Luo S, Chen L, Yan S, Gao Y, Wu G (2011) Deep into software managed coherence on SCC, July 2011
3. Lankes S, Reble P, Clauss C, Sinnen O (2011) The path to Metal SVM: shared virtual memory for the SCC. In: Proceedings of the 4th many-core applications research community (MARC) symposium
4. Prescher T, Rotta R, Nolte J (2011) Flexible sharing and replication mechanisms for hybrid memory architectures. In: Proceedings of the 4th many-core applications research community (MARC) symposium, pp 67–72
5. Kim J, Seo S, Lee J (2011) An efficient software shared virtual memory for the single-chip cloud computer. In: Proceedings of the 2nd Asia-Pacific workshop on systems. ACM, pp 4:1–4:5
6. Sivaramakrishnan KC, Ziarek L, Jagannathan S (2012) A coherent and managed runtime for ML on the SCC. In: Proceedings of the 7th many-core applications research community (MARC) symposium
7. Chapman K, Hussein A, Hosking AL (2011) X10 on the single-chip cloud computer: porting and preliminary performance. In: Proceedings of the 2011 ACM SIGPLAN X10 workshop. ACM, pp 7:1–7:8
8. Peter S, Roscoe T, Baumann A (2013) TN05—Barrelfish on the Intel single-chip cloud computer. Technical report

Review and Comparison of Mobile Payment Protocol

Pensri Pukkasenung and Roongroj Chokngamwong

Abstract Mobile phones are getting smarter and people have been using them for many different proposes. Recently, more and more people have begun using their mobile phones as a method of payment for online shopping and banking. Mobile payments have become easier than ever. Present security issues of mobile payments, however, still require improvement. This paper aims to summarize the idea of mobile payments and analyze the research of existing secure mobile payment protocols by using MPPS (Mobile Payment Protocol Security) framework. As a result, this paper will give researchers tools to standardize current protocol and share new developments.

Keywords Mobile payment · Payment protocol · Secure mobile payment

1 Introduction

Mobile devices have become a popular method for businesses in the digital world because of their convenience for payments of goods and services. The payers can access the payment system via web browsers or applications on mobile devices. More people nowadays are willing to pay for goods or services using their mobile devices. Gartner Inc., the world's leading information technology, reports that the market worth of worldwide mobile payment transactions grew to $235 billion in 2013 and will reach $721 billion by 2017 [1]. Thrive Analytics surveyed the

P. Pukkasenung (✉)
Faculty of Science and Technology, Rajabhat Rajanagarindra University,
422 Marupong Road, Muang Chachoengsao, Thailand
e-mail: pensri.puk@csit.rru.ac.th

P. Pukkasenung · R. Chokngamwong
Faculty of Information Science and Technology, Mahanakorn University of Technology,
140 Cheumsaman Road, Nongchok Bangkok, Thailand
e-mail: roongroj@mut.ac.th

© Springer Science+Business Media Singapore 2016   11
J.J.(Jong Hyuk) Park et al. (eds.), *Advances in Parallel and Distributed Computing
and Ubiquitous Services*, Lecture Notes in Electrical Engineering 368,
DOI 10.1007/978-981-10-0068-3_2

consumers in Asia-Pacific region and the results showed that there are about 800 million people who have used mobile phones as of June 2014 [2]. Thrive Analytics also found that 46 % haven't used a mobile phone to pay for goods and services because they concern about security and privacy [2]. Thus, the study concluded that the mobile payments have both advantages and disadvantages. The researchers are trying to find ways to deal with privacy and security issues by designing a protocol for mobile payments to be more effective and secure.

This paper analyzed the mobile payment protocols dating back 10 years in three aspects: methodology, security and performance. The structure of the paper is organized as follows. Section 2 provides an overview and the background of mobile payments. Section 3 classifies the technology of mobile payment systems. Section 4 presents the properties of security and cryptographic concept. Section 5 analyzes the existing secure mobile payment protocols. Section 6 concludes the paper.

2 Background and Related Work

This section provides the background and related works of the mobile payment.

2.1 Primitive Payment Transaction

Conceptually, the primitive mobile payment is composed of three basic steps [3, 4]: *Payment*—Client makes a payment to the merchant, *Value Subtraction*—Client requests to the payment gateway for his debit, and *Value Claim*—Merchant requests to the payment gateway to credit transaction amount into his account.

2.2 Components of Mobile Payment

We analyze the components of mobile payments from the existing researches related to mobile payment protocols. Fun, Beng and Razali stated that the components of mobile payment scheme consist of seven main actors: Financial Service Providers (FSPs), Payment Service providers (PSPs), Payee, Payer, Mobile Network Operator (MNOs), Device Manufacturers, and Regulators [5]. However, Fun, Beng, Roslan and Habeeb stated that mobile payment protocols are composed of five principals which include client, merchant, issuer (client's financial institution), acquirer (merchant's financial institution) and payment gateway (PG) [3]. Kungpisdan, Srinivasan and Le also defined that five parties on mobile payment protocols are client, merchant, payment gateway, issuer and acquirer [4, 6]. Singh and Shahazad stated that the components of mobile payment protocol consist of three participants: payee, payer and financial institution [7]. McKitterick and

Dowling stated that the components of mobile payment protocols are composed of four parts: customer, merchant, payment service provider and trust third party (TTP) [8].

The number of components mentioned above by researchers is different due to the design of payment protocols. However, we conclude that the components of mobile payment protocols, in general, consist of only three main parts: buyer, payment channel and seller.

2.3   Mobile Payment Procedure

Type of payments based on location

- *Remote Transactions*: These transactions are conducted regardless of the user's location. Location distances don't limit the users.
- *Proximity/Local Transactions*: These transactions are where the device communicates locally to perform close proximity payments. This involves the use of short range messaging protocol such as Bluetooth infrared, RFID and contactless chips to pay for goods and services in short distances.

Type of payments based on value

- *Micro-Payments*: These are low value payments less than US$1 [5].
- *Macro-Payments*: These are large value payments more than US$10 [5].

Type of payments based on charging method

- *Post-paid*: This is the most common payment method used in e-commerce transactions today. This consists of account-based and token-based method. Account-based method is used by banks, and the credit card industry. Consumers with a bank account or credit card can pay using the account-based method [7]. Token-based method is the charge method for goods and service such as e-money, e-wallet by mobile network operator [9, 10].
- *Pre-paid*: This is the most common charging method used by mobile network operators as well as third-party service providers. This method can only be used by consumers capable of paying immediately.

3   Technology of Mobile Payment

We studied and assessed technologies in mobile payment systems from the existing researches as described below [11].

- *SMS*—Short Messaging Service is a text messaging service used to send and receive short text messages. The maximum length of messages is less than 160 alphanumeric characters, to and from mobile phones.
- *WAP*—Wireless Application Protocol is a technology which provides a mechanism for displaying internet information on a mobile phone.
- *NFC*—Near Field Communication is the communication between contactless smart cards and mobile phones.
- *RFID*—Radio Frequency Identification is a method of identifying an item wirelessly using radio waves
- *Smart Card*—Smart cards and plastic cards normally appear in the same shape as credit cards are embedded with a chip or microprocessor that can handle and store 10–100 times more information than traditional magnetic-stripe cards [12].
- *Internet*—The internet is a publicly accessible, globally interconnected network. It uses the internet protocol to enable the exchanging and sharing of data among computers in the network
- *USSD*—Unstructured Supplementary Services Data is a mechanism of transmitting information via a GSM network. Unlike SMS, it offers a real-time connection during a session
- *IVR*—Interactive Voice Response is a telephony technology where the users can interact with the database of a system without any human interaction
- *Magnetic*—Data is stored in a magnetic stripe on a plastic card. It is read by swiping the card in a magnetic card reader.

4 Security of Mobile Payment

This section presents security properties, and cryptographic techniques.

4.1 Security Properties

A secure mobile payment system must have the following properties [13].

- *Confidentiality*—The system must ensure that private or confidential information will not be made available or disclosed to unauthorized individuals.
- *Integrity*—The system must ensure that only authorized parties are able to modify computer system assets and transmitted information.
- *Authentication*—The system must ensure that the origin of a message is correctly identified, with an assurance that the identity is not false.
- *Non-repudiation*—The system must ensure that the user cannot deny that he/she has performed a transaction and he/she must provide proof if such a situation occurs.

- *Availability*—The system must be accessible for authorized users at any time.
- *Authorization*—The system must verify if the user is allowed to make the requested transaction.

4.2 Cryptography Concept

Cryptography is a technique used to secure data protection from the hacker, which can be classified into the following three groups:

- *Symmetric Key Cryptography*—It is the encryption methods in which both the sender and receiver share the same key. The algorithms,in general, consist of DES (Data Encryption Standard), 3DES (Triple DES) and AES (Advance Encryption Standard)
- *Asymmetric Key Cryptography*—It is also known as public key cryptography, a class of cryptographic algorithms which requires two separate keys. One key is secret and the other key is public.The algorithms are RSA (Rivest, Shamir and Adleman) and ECC (Elliptic Curve Cryptography).
- *Hash Function*—It is a public one-way function that maps a message of any length into a fixed-length, which serves as the authenticator. A variety of ways of a hash code can be used to provide message authentication.

5 Analysis of Existing Secure Mobile Payment Protocols

We analyzed the existing researches on 11 secure mobile payment protocols that focus on lightweight protocol and high level of security. Bellare and Wang [14] designed the SET protocol (Secure Electronic Transfer Protocol) in 1996. This protocol is using a cryptographic technique by using public key and digital signature to protect information on mobile payment via a credit card that gives three important properties of information security: confidentiality, integrity and authorization. Bellare and Garay [15] designed the iKP protocol (i-Key-Protocol) in 2000 that is adjusted from the SET protocol by using pair "i". If it is high, it shows a high level of security. This protocol provided the properties of security similar to the SET protocol. Kungpisdan and Srinivasan [16] designed the KSL protocol (Kungpisdan Logic) in 2003 which focuses on client processing for decreasing the computational cost on the mobile wireless network. The protocol applied a symmetric key cryptography. The comparison shows that it has better performance over the SET and iKP protocols and also provides the non-repudiation property. Kungpisdan et al. [4] developed the Kungpisdan Protocol (Account-based Mobile Payment) in 2004 that is improved from KSL protocol by using symmetric key for all the parties. This protocol creates a secret shared key between two parties which support high level of four security properties: confidentiality, integrity, authentication and non-repudiation. The performance, when compared with the SET and iKP protocol, showed that the computation time at the client is relatively faster.

Fun et al. [17] designed the LMPP protocol (Lightweight Mobile Payment Protocol) in 2008. This protocol is using only the symmetric key but the performance is better than the SET, iKP and Kungpisdan [16] protocols. Shedid [18] adjusted the MSET Protocol (Modified SET Protocol) in 2010 by decreasing the number of operational cryptographic for increasing the performance. Dizaj et al. [19] designed the MPCP2 Protocol (Mobile Pay Center Protocol 2) in 2011 for decreasing the number of cryptographic operations between all engaging parties. By using symmetric cryptography all parties exchange key offline by Diffie-Hellman method. When compared with the SET, iKP, KSL and Kungpisdan protocols, the performance showed that the number of operation at the client is less than the number of operation of the other protocols. Isaac and Zeadally [20] designed PCMS Protocol (Payment Centric Model Using Symmetric Cryptography) in 2012. The protocol focuses on Payment gateway centric model. All parties must connect via the payment gateway for authorization.

Sekhar and Sarvabhatla [21] designed the SLMPP Protocol (Secure Lightweight Mobile Payment Protocol) in 2012. This protocol focuses on end-to-end encryption by using symmetric key cryptography in order to decrease the number of operation at the client side. The comparison with the SET, iKP and Kungpisdan protocols found that this protocol has less number of operations. The authors concluded that this protocol is suitable for mobile wireless network. Tripathai [22] designed the LPMP Protocol (Lightweight Protocol For Mobile Payment) in 2012 focusing on the number of cryptographic operations. It is compared with the SET, iKP, KSL and MSET protocols, and found that the LPMP use only the cryptographic operations on the client side which all processes are less than the others. Auala and Arora [23] designed the SAMPP Protocol (Secure Account-based Mobile Payment Protocol) in 2013 by using asymmetric key and digital signature. The authentication technique is using a multifactor authentication with a biometric and private key. The performance is better when compared with the SET and iKP protocols.

The analyses of the relationship between all secure mobile payment protocols from the past to present showed that almost all protocols are compared in performance with SET and iKP. Subordinates of SET and iKP are Kungpisdan, KSL, LMPP and MSET. The relationship of the secure mobile payments protocols from the past 10 years is depicted in Fig. 1a. The original protocol, SET, was formed in 1996 and the latest protocol, SAMPP, was formed in 2013. Security protocols can be divided into three aspects: methodology, security and performance. These three aspects are key factors to the success of secure mobile payment protocol and are the core of research on mobile payment security. The concept of MPPS framework is depicted in Fig. 1b.

The detailed analysis of secure mobile payment protocol is as follows:

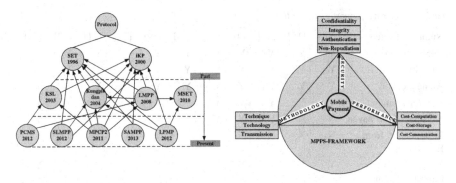

Fig. 1 The concept of secure mobile payment. **a** The pairing of the protocol. **b** Core framework of secure mobile payment protocol

5.1 Methodology Aspect

Secure mobile payment protocols such as SET, iKP, KSL and SAMPP use the asymmetric key cryptography technique to support security. The others use symmetric key cryptography. The processes of encryption have the following objectives:

- *Encryption/Decryption*—This ensures that the data is confidential and is encrypted secretly and privately.
- *Hash Function*—This ensures that the data is sent correctly and the sent data matches the original data. HMAC (Hash Message Authentication Code): provides an easy mechanism for verifying both user authenticity and that a message hasn't been tampered with of message; it protects the integrity and the authenticity of the message.
- *Key generation*—This ensures the non-repudiation property by confirming the corresponding data before beginning a transaction order to prevent disclaimers.

5.2 Security Aspect

After analyzing 11 protocols of mobile payments, we found that almost all protocols support features of security in four key areas: confidentiality, integrity, authentication and non-repudiation. But, two protocols SET and iKP do not support non-repudiation. Moreover, the protocols KSL, LMPP, MSET, MPCP2, SLMPP, LPMP and SAMPP provided all privacy properties that the others could not. The security properties and features of the different protocols are summarized in Table 1.

Table 1 Security properties of protocols

Number of reference	[14]	[15]	[16]	[4]	[17]	[18]	[19]	[20]	[21]	[22]	[23]
Confidentiality	Y	Y	Y	Y	Y	Y	Y	Y	Y	Y	Y
Integrity	Y	Y	Y	Y	Y	Y	Y	Y	Y	Y	Y
Authentication	Y	Y	Y	Y	Y	Y	Y	Y	Y	Y	Y
Non-repudiation	Y	Y	Y	Y	Y	Y	Y	Y	Y	Y	Y
Id protect from payee	N	N	Y	Y	Y	Y	Y	Y	Y	Y	Y
Id protection from Eavesdropper	N	N	N	N	Y	Y	Y	–	Y	Y	Y
Transaction privacy protection from eavesdropper	Y	Y	Y	Y	Y	Y	Y	–	Y	Y	Y
Transaction privacy protection from TTP or related financial	N	N	N	N	Y	Y	Y	–	Y	Y	Y

Table 2 Number of Cryptographic Operation of Protocol

Number of reference		[14]	[15]	[16]	[4]	[17]	[18]	[19]	[20]	[21]	[22]	[23]
Public Encryption	C	1	1									1
	M	1		1								
	PG	1		1								
Public Decryption	C											
	M	1										1
	PG	2	1	1								
Signatures Generations	C	1	1									
	M	3	3	1								
	PG	1	1	1								
Signature Verifications	C	2	3									
	M	2	2	1								
	PG	1	2	1								
Symmetric Encrypt Decrypt	C	2		3	4	5	11	4	3	4	6	3
	M			4	5	6		4	4	4		6
	PG	1			2			3	3	3		2
Hash Function	C	3	2	3	2		6	1	2	1	5	
	M	2	4	3		2		3		3		
	PG		1			1						
Key Hash	C			2	2			1	2	1		
	M		1		2			1	2	1		
	PG				1				1			
Key Generation	C			2	2			2	2	2	1	1
	M			1	1			1	1	1		2
	PG				1			1	1	1		1
Total		24	22	26	22	14	17	21	21	21	12	17

5.3 Peformance Aspect

Protocol's performance is analyzed by counting the number of operations needed for encoding and decoding. This includes operations related to data transmission between three parts. Table 2 summarizes the number of cryptographic operations which consist of public encryption–decryption, signature verifications, symmetric key encryption-decryption, a hash function, keyed-hash function and key generations. The researchers presented secure mobile payment protocols providing a high level of security and low computation, cost and power.

6 Conclusion

This paper gives an overview of mobile payments and analyzes the existing secure mobile payment protocol over the past 10 years. All protocol schemes focus on reducing the use of resources in the mobile process by cryptographic concept. Each researcher tried to design and modify the process of the protocol so the message is short and lightweight. All protocols provided four main security properties: confidentiality, integrity, authentication, and non-repudiation. As a conclusion, to discover the best secure mobile payment protocol, the protocol standard must be the same all over the world and the communities and industries must be adopting the standard.

References

1. Gartner.com (2013) Gartner Says worldwide mobile payment transaction value to surpass $235 billion in 2013. http://www.gartner.com/newsroom/id/2504915
2. Richter F Consumers wary of mobile payment security. http://www.statistica.com
3. Fun TS, Beng LY, Roslan R, Habeeb HS (2008) Privacy in new mobile payment protocol. World Acad Sci Eng Technol 2:198–202
4. Kungpisdan S, Srinivasan B, Le PD (2004) A secure account-based mobile payment protocol In: Proceedings of the international conference on information technology: coding and computing (ITCC 2004)
5. Fun TS, Beng LY, Razali MN (2013) Review of mobile macro-payments schemes. J Adv Comput Netw 1(4)
6. Kungpisdan S, Srinivasan B, Le PD (2003) Lightweight mobile credit-card payment protocol. Lect Notes Comput Sci 2904:295–230
7. Singh A, Shahazad KS (2012) A review: secure payment system for electronic transaction. Int J Adv Res Comput Sci Softw Eng 2(3)
8. McKitterick D, Dowlin J State of the art review of mobile payment technology. https://www.scss.tcd.ie/publications/tech-reports/reports.03/TCD-CS-2003-24.pdf
9. Ahamad SS, Udgata SK, Nair M (2014) A secure lightweight and scalable mobile payment framework. In: FICTA 2013. Advances in intelligent system and computing, vol 247. Springer International Publishing, Switzerland

10. Ferreira C, Dahab R (1998) A scheme for analyzing electronic payment systems. In: Computer security applications conference, proceedings. 14th Annual, 1998
11. Mathew M, Balakrishnan N, Pratheeba S (2010) A study on the success potential of multiple mobile payment technologies. In: Technology management for global economic growth (PICMET), Proceedings of PICMET '10
12. Smart Card Alliance (2008) Proximity mobile payments business scenario: research report on stakeholder perspectives
13. Computer Fraud & Security (2007) Analysis of mobile payment security measures and different standards
14. Li Y, Wang Y Secure electronic transaction. http://people.dsv.su.se/~matei/courses/IK2001SJE/li-wang_SET.pdf
15. Bellare M, Garay JA (2000) Design implementation, and deployment of the ikp secure electronic payment system. IEEE J Sel Areas Commun 18(4)
16. Kungpisdan S, Srinivasan B, Le PD (2003) Lightweight mobile credit-card payment protocol. Lect Notes Comput Sci 2904:295–308
17. Fun TS, Beng LY, Likoh J, Roslan R (2008) A lightweight and private mobile payment protocol by using mobile network operator. In: Proceedings of the international conference on computer and communication engineering 2008 May 13–15, Kuala Lumpur, Malaysia, 2008
18. Shedid SM (2010) Modified SET protocol for mobile payment. Proc Int Conf J Comput Sci Netw Secur 10(7):289–295
19. Alizadeh Dizaj MV, Moghaddam RA, Momenebellah S (2011) New mobile payment protocol: Mobile Pay Center Protocol 2 (MPCP2) by using new key agreement protocol: VAM. In: 3rd international conference on electronics computer technology (ICECT)
20. Isaac JT, Zeadally S (2012) An anonymous secure payment protocol in a payment gateway centric model. In: The 9th international conference on mobile web information system (MobiWIS). Elsevier
21. Sekhar VC, Sarvabhatla M (2012) Secure lightweight mobile payment protocol using symmetric key techniques. In: International conference on computer communication and informatics (ICCCI), pp 1–6, 10–12 Jan 2012
22. Tripathi DM, Ojha A (2012) LPMP: an efficient lightweight protocol for mobile payment. In: 3rd national conference on emerging trends and applications in computer science (NCETACS)
23. Auala PS, Arora H (2013) A secure account based mobile payment protocol with public key cryptography and biometric characteristics. In: International journal of science and research (IJSR), vol 2(3), India online ISSN: 2319-7064

POFOX: Towards Controlling the Protocol Oblivious Forwarding Network

Xiaodong Tan, Shan Zou, Haoran Guo and Ye Tian

Abstract Protocol Oblivious Forwarding (POF) is a recently proposed technology that enables a protocol independent data plane under the context of Software-Defined Networking (SDN). In this paper, we present POFOX, a SDN controller for POF. POFOX employs the full potentials of POF devices by allowing a protocol oblivious data plane, and it provides a simple programming model similar to POX. Based on POFOX, we construct a network testbed, and experimentally illustrate that POFOX can effectively manage the POF network, and provide the controlling functionality with high performances.

Keywords Software-Defined Networking (SDN) · OpenFlow · Protocol Oblivious Forwarding (POF) · Network Testbed

1 Introduction

Software-Defined Networking (SDN [1]) provides a dynamic and cost-effective way for managing the network. However, the OpenFlow 1.x specification [2], which is currently the de facto standard of SDN, becomes more and more complicated by supporting more and more protocols. To overcome this problem, technologies such Protocol Independent Forwarding (OF-PI [3] and P4 [4]) and

X. Tan · S. Zou · H. Guo · Y. Tian (✉)
School of Computer Science and Technology, University of Science
and Technology of China, Hefei Anhui 230026, China
e-mail: yetian@ustc.edu.cn

X. Tan
e-mail: xdtan@mail.ustc.edu.cn

S. Zou
e-mail: jelly33@mail.ustc.edu.cn

H. Guo
e-mail: hrguo@mail.ustc.edu.cn

© Springer Science+Business Media Singapore 2016   21
J.J.(Jong Hyuk) Park et al. (eds.), *Advances in Parallel and Distributed Computing
and Ubiquitous Services*, Lecture Notes in Electrical Engineering 368,
DOI 10.1007/978-981-10-0068-3_3

Protocol Oblivious Forwarding (POF [5]) are proposed very recently. In this paper, we focus on POF, which was initially proposed by Huawei Technologies [6]. In POF, a data plane forwarding device does not need to know the header structure of the protocol packet that it is supposed to forward in advance, but can be configured by the control plane with the offsets in the configuration phase. At run-time, the device extracts the packet field by the offsets, and looks up the flow table and executes the relevant instructions. In this way, the forwarding device can easily support new protocols without modifying its hardware as well as the wire protocol between controller and switch (e.g., OpenFlow [7]).

Unlike OpenFlow which defines the supported protocols in its specification, POF is entirely oblivious on the data plane protocols such as Ethernet, IP, TCP, etc. In POF, when a forwarding device receives a packet, it extracts the packet's header layer by layer, and each layer has multiple flow tables. A flow table not only processes the current layer, but also parses header of the next layer.

Currently, Huawei only releases the prototypes of POF switch and controller as Linux codes [8]. The major problem for Huawei's POF controller is that it can only configure the flow tables of the forwarding devices manually, but do not support any programming model. However, the most important and attractive feature of SDN is its programmability. In this work, we present POFOX, a SDN controller for POF. Unlike the Huawei controller, POFOX is highly flexible and enables users to manage the POF devices with Python programs. In the design and implementation of POFOX, we have borrowed the programming model and the basic architecture of POX [9], a very popular OpenFlow controller, and re-design and re-code the key components of the controller such as the communication engine and the topology discovery component to support the POF devices. We have also constructed the first programmable POF network testbed with POFOX, and examine the network's performance with real-world experiments.

The remainder part of this paper is organized as follows. We describe the frame structure and fundamental component of the POFOX in Sect. 2. In Sect. 3, we evaluate the performance of POFOX by the experiments on the network testbed to show the benefits of our design. We conclude and discuss the future work in Sect. 4.

2 POFOX Design

We demonstrate the basic design of POFOX in Fig. 1. As shown in the figure, the fundamental component of POFOX is the communication engine, it is responsible for contacting with related switches by sending and receiving the POF messages. Based on this component, we realize a rich set of applications to meet different goals, such as the topology discovery and spanning tree protocol component. We will describe the components respectively in the following subsections.

Fig. 1 Frame structure of POFOX

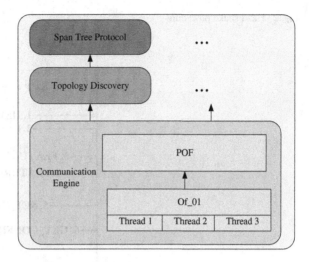

2.1 Communication Engine

In POF, the POF switch only needs to forward the packets under the control of the POFOX. Thus, the most important thing POFOX needs to do is to communicate with the switch with interactive messages. The messages are used to establish the connection, query the information of the flow tables and change the status of the switch, etc. In this component, we define the data structure of these messages in detail and create the sockets to listen to the messages from the switches, when POFOX receives the messages, it will unpack the packets and forward to other components. Instead, it also packs the data when sends messages to the switches.

The procedure of building the connection between POFOX and POF switch called "hand shake" is shown in Fig. 2. Initially, the switches join the network by setting up a TCP connection with POFOX, it sends "HELLO" and POFOX replies with the same message. Then, the "FEATURE" tells POFOX about the features of the switches and the "PORT" is about the physical port status of the switches. The process of the other messages is in a similar way.

2.2 Topology Discovery

The POFOX takes charge of the behaviors of the POF switch, thus it is crucial to discover the topology of the switches. This component sends out probing messages, which conform to the vendor-neutral Link Layer Discovery Protocol (LLDP [10]),

Fig. 2 Hand shake procedure

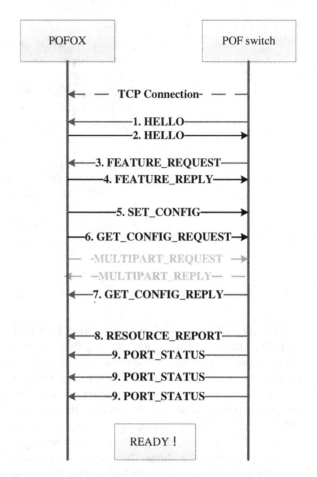

to the neighbors of each switch at a fixed interval. Then it receives bounced back probing messages and discovers the topology of the network after unpacking the LLDP packets. The algorithm of topology discovery is shown in Algorithm 1.

Algorithm 1. Topology discovery algorithm of POF network

Input: *cycle_time* : the time interval of the POFOX updates the topology information;

Define: *this_cycle* = []: list of the interfaces which are remaining to send the probing messages in this cycle;
next_cycle = []: list of the interfaces have sent the probing messages in this cycle;
tree_dict = {}: a dictionary saves the topology information of the POF switches;

Steps:

```
for every cycle_time seconds do
            if a POF switch connects or disconnects the
POFOX then
            update this_cycle and next_cycle;
          end if
len = len(this_cycle) + len (next_cycle)
for every cycle_time / len seconds do
          if len(this_cycle) == 0 then
            this_cycle = next_cycle;
            next_cycle = [];
          end if
          p1 = this_cycle.pop[0];
          next_cycle.add(p1);
          POFOX tells POF switch s1 to send a probing
message from p1;
          if POF switch s2 receives a probing message
from s1 by p2 then
            forward the probing message to POFOX
          end if
          if POFOX receives a probing message do
            link = (s1, p1, s2, p2);
            tree_dict.add(link);
          end if
        end for
end for
```

When a POF switch connects or disconnects POFOX, *this_cycle* and *next_cycle* updates the ports information. For a fixed interval *cycle_time*, POFOX will send the probing messages from the ports in the *this_cycle*, and the ports have sent the messages will be added in the *next_cycle*, and if another switch receives the probing messages, it will match the flow table and forward the messages to POFOX. POFOX will discover a *link* between *s1* and *s2* after parsing the packets. After this cycle, POFOX gathers all the links and detects the topology of the switches. We will show the experiments to examine the functionality of this module.

2.3 Spanning Tree Protocol

According to the topology information, we can find that there may be a loop between the switches. As a result, the switches will forward the packets again and again until the Time to Live (TTL), which is our defined field rather than the filed in IP packets, decreases to 0. It will generate the broadcast storm and take up a lot of network bandwidth, even make the network completely paralyzed. In order to avoid this condition, we construct a spanning tree of the network topology, and then disable flooding on switch ports that are not in the tree. The realization of this component is shown as Algorithm 2.

Algorithm 2. Spanning Tree Protocol algorithm of POF network

```
Input: tree_dict: a dictionary about the topology
information of the POF switches;
switches: a set contains all of POF switches connect
the POFOX;

Define: done = {}: POF switches that have been visited;
tree_port = {}: a set contains the ports which are able
to flood;

Steps:
for s1 in switches do
            if s1 not in done then
              done.add(s1)
            else
              continue;
            end if
            for s2 in switches do
              if there is a link between s1 and s2 in
tree_dict then
                done.add(s2)
                tree_port.add(s1.port, s2.port)
              end if
            end for
end for
for s in switches do
            for port in s do
              if port in tree_port then
                s.port.is_flood = TRUE
              else
                s.port.is_flood = FALSE
              end if
            end for
end for
```

We get the topology information *tree_dict* and the set of all the POF switchs *swithes*, then use *done* to store the POF switches which have been visited and *tree_port* to save the ports could flood. We visit the switch in *switches* one by one, if a switch *s1* is in the *done* means that it has been visited, and we need to visit the next one. Otherwise, we visit the remaining switch in *switches* to check whether there is a connection between *s1* and another switch. If another switch *s2* exists, we add the ends of this link *s1.port* and *s2.port* to *tree_port*. Until the *switches* is empty or the *done* contains all the POF switches, we turn on the function of the flooding of all the ports in *tree_port*. Finally, there comes to be a spanning tree to avoid the broadcast storm. We also achieve an experiment to describe the performance of this component.

3 POF Network Testbed

3.1 Devices

To evaluate both the functionality and performance of our design, we build a real-world POF network testbed. Figure 3 shows the topology of the testbed, we run POFOX on a general PC which configuration is Intel Core i7, 1T hard drive and 8G of memory, and we run POF switch on a PC which is shown in Fig. 4 and has ten 1Gbps NICs. The switches communicate with each other by a gigabit Ethernet switch, and POFOX also connects the Ethernet switch.

Fig. 3 Logical topology of the POF network

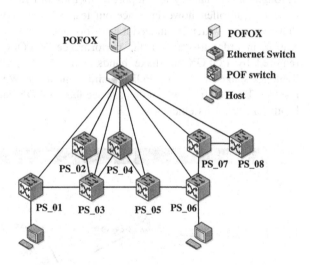

Fig. 4 POF network testbed

3.2 Functionality Test

In order to examine the functionality of POFOX, we run the Topology Discovery module to describe the switches topology of the POF network testbed. In order to more intuitive description, we use a third party tool poxdesk [11] to show the result. As is shown in Fig. 5, there are eight nodes and each node represents for a switch. Compare to the Fig. 3, we can easily find that the topology we discover is as same as the real topology. We also accomplish the experiment many times by different number of switches, we don't show the result here because of the space limitation.

3.3 Performance Test

3.3.1 Connection Time

Throughput and latency are important metrics to evaluate the performance of the network controller, however, since our testbed is currently composed of software POF switches, which are indeed the performance bottleneck of the testbed. For this reason, in order to estimate the performance of POFOX, we measure the time required by POFOX to shake hands with different number of POF switches. Switches shake hands with POFOX simultaneously. We present the measurement result in Table 1, from which one can see that POFOX can handle the simultaneous handshake requests efficiently.

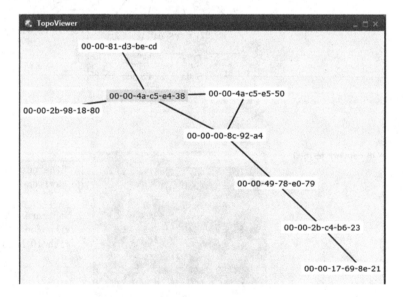

Fig. 5 Topology of the POF switches in testbed

Table 1 Connection time of different number of switches

Num. of switches	1	2	3	4	5	6	7	8
Connection time/s	0.33	0.64	0.89	1.15	1.42	1.80	2.02	2.21

Table 2 The number of packets of two situations

POF switch	With STP module	Without STP module
PS_01	100	500
PC_03	100	500
PC_04	100	600
PC_05	100	600
PC_06	100	500

3.3.2 Avoiding Broadcast Storm

On condition that POF is oblivious to the protocol of the packets, we can create a new protocol packet by Ostinato [12] to test the performance of this component. We set the Ethernet type to 0×0989, and set the TTL value to 10. Meanwhile, we implement a L2 MAC Learning application on the controller. When the packets arrive in a switch, the switch decrements the current TTL value in the packet before forwarding it by the instructions of the flow table, and the packet could not forward to other switches when TTL comes to 0 or it arrives at the destination. Then, we send 100 packets on PC_01 to PC_02 and use the Wireshark to capture the packets on the switches in the link. In Table 2, we compare two situations of whether running this component. It could easily find that the number of the packets is many times than without running this module. Thus, this component reduces the transmission of the redundant packet and improves the stability of the network effectively.

4 Conclusion and Future Work

Programmability and flexibility make SDN a popular choice for different networking scenarios today [13]. However, Huawei's controller only can configure the switch manually and ignores these vital features. Instead, in POFOX, we make full use of the programmability of POFOX and achieve rich networking functions, we can also change the control plane functionality by modifying the components or adding new modules to meet different goals flexibly.

Many other tasks and research work lay ahead: we will exploit parallelism [14] and achieve massive performance improvement both of POFOX and POF switch. The performance improvement will have a positive impact on our subsequent research. Although we are still at an early stage and have a lot of hard work remains to be done, we believe that it is significant to the research of the future network.

Acknowledgements This work was supported by the sub task of the Strategic Priority Research Program of the Chinese Academy of Sciences under Grant No. XDA06011202 and the National Natural Science Foundation of China under the Grant No. 61202405.

References

1. Casado M, Freedman MJ, Pettit J, Luo J, McKeown N, Shenker S (2007) Ethane: taking control of the enterprise. In: Proceedings of ACM SIGCOMM'07, Kyoto, Japan
2. OpenFlow Specification. https://www.opennetworking.org/technical-communities/areas/specification
3. OF-PI: a protocol independent layer (2014) version 1.1, ONF TR-505
4. Bossharty P, Daly D, Gibb G et al (2014) "P4: programming protocol-independent packet processors," ACM SIGCOMM Comput Commun Rev, 44(3):87–95
5. Song H (2013) "Protocol-oblivious forwarding: Unleash the power of SDN through a future-proof forwarding plane". In: Proceedings of ACM SIGCOMM Workshop on HotSDN, Hong Kong
6. Huawei Technologies. http://www.huawei.com/
7. McKeown N, Anderson T, Balakrishnan H, Parulkar G, Peterson L, Rexford J, Shenker S, Turner J (2008) OpenFlow: Enabling innovation in campus networks. ACM SIGCOMM Comput Commun Rev 38(2):69–74
8. POF. http://www.poforwarding.org/
9. POX. http://www.noxrepo.org/pox/about-pox/
10. LLDP. http://en.wikipedia.org/wiki/Link_Layer_Discovery_Protocol
11. Poxdesk. https://github.com/MurphyMc/poxdesk/wiki/Getting-Started
12. Ostinato. https://code.google.com/p/ostinato/wiki/UserGuide
13. Zhang C, Cui Y, Tang H, Wu J (2015) "State-of-the-art survey on software-defined networking (SDN)," Ruan Jian Xue Bao/J Soft (in Chinese), 26(1):62–81
14. Cai Z, Cox A, Ng T (2010) "Maestro: a system for scalable OpenFlow control," Tech Rep, TR10-08, Rice University

An Experimental Study on Social Regularization with User Interest Similarity

Zhiqi Zhang and Hong Shen

Abstract Recommender Systems (RS) is widely employed in information retrieval in social networks due to the prevalence of social networking services. Since the matrix factorization (MF) model has a good expandability, social information is easy to be integrated into the model. In general, researchers convert social information to social regularization. Moreover, similarity function is the key in social regularization to constrain the MF objective function. Previous researchers defined the similarity of users' rating behavior on the same items as users' interest in similarity. However, they neglected two problems: First, the friendship is a superficial social network that cannot reflect the intimacy among users. Second, the superficial social network generally cannot represent users' interest in similarity. Recently, researchers have found that both the number of co-friends and friends sub-graph improve users' interest in similarity, but they do not give a mathematical definition. In this paper, we use these two factors to design two new similarity functions. To use them in the MF-based RS, we come up with two kinds of social regularization for each similarity function. Compared with previous social regularization, our methods can more precisely explain users' interest similarity. The experimental analysis on a large dataset shows that our approaches improve the performance of the state-of-the-art social recommendation model.

Keywords Recommender system · Social network · Matrix factorization · Social regularization

Z. Zhang (✉)
School of Computer and Information Technology,
Beijing Jiaotong University, Beijing, China
e-mail: zhiqizhang@bjtu.edu.cn

H. Shen
School of Information Science and Technology,
Sun Yat-Sen University, Guangzhou, China
e-mail: hshen@bjtu.edu.cn

H. Shen
School of Computer Science, University of Adelaide, Adelaide, China

© Springer Science+Business Media Singapore 2016
J.J.(Jong Hyuk) Park et al. (eds.), *Advances in Parallel and Distributed Computing and Ubiquitous Services*, Lecture Notes in Electrical Engineering 368,
DOI 10.1007/978-981-10-0068-3_4

1 Introduction

Due to the prevalence of social networking services, social-information based recommender systems (RS) have become an emerging research topic.

Social information mainly has two instructions: trust relationship and friendship. However, [1] indicates that friendship is better than trust relationship and friendship can really reflect our offline life. In order to close to the real life, we use friendship. Note that the terms friendship and social network are used as synonyms throughout this paper. In general, most social RS are based on matrix factorization (MF) and they convert social information to social regularization [1–7]. Typically, similarity function is the main influence gene in regularization. Previously, there are two main similarity functions: Vector Space Similarity (VSS) and Pearson Correlation Coefficient (PCC). However, these functions still have several inherent limitations and weaknesses that need to be addressed.

First of all, due to the data sparsity problem, the number of co-items two users share is quite small. If target user u_i and his friend u_f have no shared item, the similarity between u_i and u_f will be zero. So it would lead to the loss of social information. If user u_i and friend u_f have only one shared item, the rating behavior on the item directly determines two users' similarity. Secondly, users' same rating custom would not explain users' interest in similarity. While some friends have the same taste with target user, others may have the opposite. Thirdly, previous researchers considered that the closeness of user u_i's friends is the same. However, it does not match to the real life and previous similarity functions only used superficial friendship (i.e. Fig. 11-1 Social Network).

As mentioned in [8], both the number of co-friends two users share and the sub-graph topology in user's friend network are two main factors that dominate users' interest in similarity. But they don't give an exactly mathematical definition of similarity. Therefore, our work is to use these two factors to define two new similarity functions, respectively. These two new similarity functions are $SFsim(i,f)$ function (uses the number of co-friends two users share as similarity value) and FRsim(i) function (uses the number of relationship between user u_i's friends). Note that the sub-graph topology in user's friend network is the relationship among friends throughout this paper. Our major findings are summarized as follows. First, SFsim(i, f) reflects the similarity between user u_i and user u_f, while FRsim(i) reflects the average similarity between user u_i and his/her friends. Second, SFsim(i, f) follows one-to-one restriction and FRsim(i) follows one-to-many restriction. Third, different similarity functions have different influences on different regularization.

The rest of the paper is organized as follows. In Sect. 2, we provide an overview of related work. Section 3 presents our new similarity functions. Section 4 details how to use our new similarity functions in social regularization terms. Experiments and discussions are given in Sect. 5. Finally, we give a conclusion.

2 Related Work

In this section, we review the popupar similarity function PCC [5].

Similarity Function. In social RS, we always need to make similar users or items into a circle. In order to compute similarity, researchers always use two methods: VSS and PCC. However, VSS does not consider that users have different rating customs. Hence, PCC is proposed to solve this problem.

$$Sim_{PCC}(i,j) = \frac{\sum_{k \in I(i) \cap I(j)} (R_{ik} - \bar{R}_i)(R_{jk} - \bar{R}_j)}{\sqrt{\sum_{k \in I(i) \cap I(j)} (R_{ik} - \bar{R}_i)^2} \cdot \sqrt{\sum_{k \in I(i) \cap I(j)} (R_{jk} - \bar{R}_j)^2}} \tag{1}$$

where $\bar{R}_i$ represents the average rate of user u_i in all his rating items. So many researchers adopt this method in social regularization. However, it only uses the similarity of rating behavior as user's interest in similarity. So, in the next section, we will discuss two new similarity functions.

3 New Similarity Functions

In this section, we define two new similarity functions and discuss how to use them in different regularization terms to constrain the MF objective function. Figure 11-2 Friend Relation shows the relationship of user u_i's friends, the fine lines represent the friendship of u_i and the thick lines represent relationship among user u_i's friends. In Fig. 11-3 Co-friends, the intersection represents shared friends between two users.

Therefore, we propose two similarity functions: *FRsim(i)* and *SFsim(i,f)*. So, we define:

$$\mathrm{FRsim(i)} = FR_i \tag{2}$$

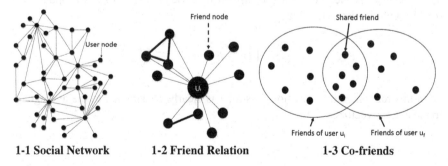

1-1 Social Network   **1-2 Friend Relation**   **1-3 Co-friends**

Fig. 1 Friendship

$$\text{SFsim}(i, f) = SF_{i,f} \tag{3}$$

where FR_i value represents the number of relationships among user u_i's friends, and $SF_{i,f}$ value represents the number of shared friends between user u_i and user u_f. For example, FR_i value in Fig. 11-2 is the number of red lines and $SF_{i,f}$ value in Fig. 11-3 Co-friends is the number of shared friend in the intersection. Besides, we employ a mapping function $f(x) = \frac{1}{1+e^{-\theta x}}$ to bound these two similarity functions range into [0, 1], where θ is an adjustable parameter. Typically, the function FRsim(i) represents average similarity between user u_i and u_i's friends, while the function SFsim(i, f) represents every two users' similarity.

4   Social Regularization

In this section, we will systematically interpret how to model social network information as regularization terms to MF-based RS.

4.1   Model 1: Friend Relationship Regularization (FRR)

In social networking services, larger value of FR_i has a positive effect on the average similarity between user u_i and his friends. Basing on this intuition, we impose our first social regularization term.

$$Reg_{FRR} = \frac{\beta}{2} \sum_{i=1}^{m} FRsim(i) \cdot \frac{\sum_{f \in \mathcal{F}_i} \| U_i - U_f \|_F^2}{N_i} \tag{4}$$

where β, $\mathcal{F}_i$ represents the set of the friends of user u_i, N_i is the size of set $\mathcal{F}_i$, and $\| \cdot \|_F^2$ denotes the Frobenius norm. However, this method will produce slight deviation. Since function FRsim(i) should act on the set of all friends' vectors not the summary distance between user u_i and u_i's friends. Therefore, we update the social regularization term Reg_{FRR+}.

$$Reg_{FRR+} = \frac{\beta}{2} \sum_{i=1}^{m} \| U_i - FRsim(i) \cdot \frac{\sum_{f \in \mathcal{F}_i} U_f}{N_i} \|_F^2 \tag{5}$$

In this regularization, function FRsim(i) is clearly constrained to the average of friends' total vectors.

4.2 Model 2: Shared Friend Regularization (SFR)

As mentioned above, FRsim(i) function is a one-to-many restriction. In general, one-to-one restriction is better than the one-to-many. Therefore, we try to propose another social regularization term.

$$Reg_{SFR} = \frac{\beta}{2} \sum_{i=1}^{m} \| U_i - \frac{\sum_{f \in \mathcal{F}_i} SFsim(i,f) \cdot U_f}{\sum_{f \in \mathcal{F}_i} SFsim(i,f)} \|_F^2 \tag{6}$$

Although $SFsim(i,f)$ function generates a one-to-one similarity value, this approach makes every constrain into averaging. It will reduce each friend's influence. Hence, in order to tackle this problem, we propose another regularization Reg_{SFR^+}.

$$Reg_{SFR^+} = \frac{\beta}{2} \sum_{i=1}^{m} \sum_{f \in \mathcal{F}_i} SFsim(i,f) \cdot \| U_i - U_f \|_F^2 \tag{7}$$

As well as SoReg model [5], this approach can indirectly minimize the distance between feature vectors U_i and U_g, when we are minimizing the distances

$$Sim(i,f) \| U_i - U_f \|_F^2 \text{ and } Sim(f,g) \| U_f - U_g \|_F^2 \tag{8}$$

Compared with SoReg model, this approach adequately considers that different friends may have different effects on the target user. Besides, function $SFsim(i,f)$ reflects the interest similarity between u_i and user u_f.

5 Experimental Analysis

In this section, we conduct several experiments to compare our four regularization terms with the state-of-the-art RS using Douban dataset.

5.1 Dataset

The data source we choosed in the experiment is Douban dataset. More specially, most friends on Douban actually know each other offline and the friendship is symmetrically. It is an ideal source for our research on social recommendation. The dataset contains three aspects: music, book, and movie. In the paper, we randomly select 80 % of the ratings to train the models.

Table 1 Statistics of friend relationship and shared friends

Statistics	Music	Book	Movie
Max. num. of FR	3815	5711	6240
Num. of no FR	2015	2230	3329
AVG. ratio of FR	0.9281	1.1652	1.0793
Max. num. of SF per user	98	113	143
Num. of no SF per user	2529	2831	4200
Num. of all users' SF	1014779	148165	203837
Sparsity of SF (%)	0.33	0.34	0.25

Table 1 gives the relationships among friends and shared friends in each dataset. In Music column, the max number of relationship among user u_i's friends is 3815 and there are 2015 users without the relationships among friends. The ratio of FR is the number of relationships among user u_i's friends in the size of user u_i's friends set. So, the large value of this ratio indicates that the friend relationship should be more closely. Besides, the max number of shared friends per user is 98 and there are 2529 users without shared friends. If every two users has shared friend, the number of shared friends will be 32,216,364, but the real number is 104,779, the sparsity is 0.33 %. Note that the sparsity of the movie is smaller than others.

5.2 Comparisons

In this section, we will compare the following different methods described in this paper as well as some baseline methods. Besides, we use MAE and RMSE metrics in Table 2.

Probabilistic matrix factorization (PMF) [9], this method only uses simple MF technique to predict missing ratings without any social network information.

Table 2 Performance comparison ($\beta = 0.01, \lambda = 0.001$)

Data	Metrics	PMF	SoReg	FRR	FRR^+	SFR	SFR^+
Music	MAE	0.5824	0.5721	0.5722	0.5726	0.5719	0.5698
	Improve	2.16 %	0.40 %				
	RMSE	0.7386	0.7322	0.7320	0.7332	0.7324	0.7290
	Improve	1.30 %	0.44 %				
Book	MAE	0.6461	0.6409	0.6417	0.6415	0.6421	0.6372
	Improve	1.38 %	0.58 %				
	RMSE	0.8189	0.8192	0.8201	0.8207	0.8208	0.8141
	Improve	0.72 %	0.62 %				
Movie	MAE	0.5940	0.5894	0.5892	0.5890	0.5899	0.5913
	Improve	0.84 %	0.07 %				
	RMSE	0.7504	0.7506	0.7503	0.7501	0.7515	0.7527
	Improve	0.04 %	0.07 %				

SoReg [5] injects social network information to basic MF. Two regularization terms are proposed: the first model they used average-based regularization, which constrains the difference between the user's taste and the average of his/her friends' tastes. The second model they used one to one regularization, which constrains the difference between one user and their friends individually. In the experiment, we only compare with individual-based regularization.

FRR and FRR^+, these are the Social Regularization methods using friend relationship. SFR and SFR^+, these are the Social Regularization methods using the number of shared friends. From the results, we can observe that our method outperforms other approaches. In music and book datasets, both SFR and SFR^+ are better than FRR and FRR^+. However in movie set, the result is a little different. From Table 1, we can find that the sparsity of shared friends in movie set is smaller than others, and the average ratio of relationship among user u_i's friends is not the worst. So, in movie dataset, both FRR and FRR^+ are better than SFR and SFR^+.

6 Conclusion

In this paper, we investigate two new similarity functions which deeply reflect the intimacy of users. The function FRsim(i) indicates the average interest similarity between user u_i and u_i's friends, while function SFsim(i, f) indicates the interest similarity of every two users. To apply these two functions to recommendation systems (RS), we propose two kinds of social regularization for each function. The experimental analysis shows that our methods are better than original social regularization. We found an interesting phenomenon that the accuracy of RS based on SFR is generally better than that based on FRR. When the density of FR_i is larger than $SF_{i,f}$, the result of RS is better on SFsim(i, f). Hence, we discover that these two similarity functions are based on the density of the social network. If the social network is very dense, the value of these two functions approximate to user interest similarity. Compared with previous similarity, our methods deeply mine the relationships among friends.

Acknowledgments This work is supported by National Science Foundation of China under its General Projects funding #61170232, Research Initiative Grant of Sun Yat-Sen University under Project 985, Australian Research Council Discovery Project DP150104871.

References

1. Hao M et al (2011) Recommender systems with social regularization. In: Proceedings of the fourth ACM international conference on web search and data mining %@ 978-1-4503-0493-1. ACM, Hong Kong, pp 287–296
2. Hao M et al (2008) SoRec: social recommendation using probabilistic matrix factorization. In: Proceedings of the 17th ACM conference on information and knowledge management %@ 978-1-59593-991-3. ACM, Napa Valley, pp 931–940

3. Jamali M, Ester M (2010) A matrix factorization technique with trust propagation for recommendation in social networks. ACM
4. Liu X, Aberer K (2013) SoCo: a social network aided context-aware recommender system. In: Proceedings of the 22nd international conference on World Wide Web. International World Wide Web Conferences Steering Committee, Rio de Janeiro, pp 781–802
5. Ma H et al (2011) Recommender systems with social regularization. Wsdm, pp 287–296
6. Meng J et al (2014) Scalable recommendation with social contextual information. IEEE Trans Knowl Data Eng 26(11):2789–2802
7. Xiwang Y, Harald S Yong L (2012) Circle-based recommendation in online social networks. In: Proceedings of the 18th ACM SIGKDD international conference on knowledge discovery and data mining %@ 978-1-4503-1462-6. ACM, Beijing, pp 1267–1275
8. Ma H (2014) On measuring social friend interest similarities in recommender systems. ACM
9. Salakhutdinov R, Mnih A (2007) Probabilistic matrix factorization. In: Neural information processing systems

Representing Higher Dimensional Arrays into Generalized Two-Dimensional Array: G2A

K.M. Azharul Hasan and Md Abu Hanif Shaikh

Abstract Two dimensional array operations are prominent for array applications because of their simplicity and good performance. But in practical applications, the array dimension is large and hence efficient design of multidimensional array operation is an important research issue. In this paper, we propose a two dimensional representation of multidimensional array. The scheme converts an n dimensional array into a two dimensional array. We design efficient algorithms for matrix-matrix addition/subtraction and multiplication using our scheme. The experimental results show that the proposed scheme outperforms the Traditional Multidimensional Array (TMA) based algorithms.

Keywords Multi dimensional array · Array operations · Matrix operation · High performance computing

1 Introduction

Array is the most common and widely used data structure. Modeling and analyzing scientific phenomena strongly requires handling large scale data of higher dimension efficiently and effectively. But handling and operating on higher dimensional array is still a challenge [1]. The cost of index computation becomes high when the number of dimension increases in a multidimensional array. Again the cache miss rate increases for higher dimensional arrays as more cache lines need to be accessed [2, 3]. That's why very few commercial compilers eventually support large number of dimensions. Thus it is important to represent a multidimensional array into an

K.M. Azharul Hasan (✉) · M.A.H. Shaikh
Computer Science and Engineering Department, Khulna University of
Engineering & Technology, Khulna 9203, Bangladesh
e-mail: az@cse.kuet.ac.bd

M.A.H. Shaikh
e-mail: hanif@kuet.ac.bd

© Springer Science+Business Media Singapore 2016   39
J.J.(Jong Hyuk) Park et al. (eds.), *Advances in Parallel and Distributed Computing
and Ubiquitous Services*, Lecture Notes in Electrical Engineering 368,
DOI 10.1007/978-981-10-0068-3_5

efficient way. In this paper we propose an array system to convert a multidimensional array into a two dimensional array. We developed the array addressing functions and hence the array operations for the proposed scheme. Since two dimensional arrays are easy to understand hence less complicated algorithms can be designed. To evaluate the proposed system, we designed efficient algorithms for matrix-matrix addition/subtraction and multiplication. Our experimental results show that the generalized two-dimensional array based algorithms shows better results than TMA based algorithms. This is because of two reasons; the index computation is easy and cache miss rate is fewer in the proposed scheme.

2 Generalized Two Dimensional Representation (G2A) Scheme

We propose an algorithm to represent an n dimensional array by a 2 dimensional array. Let $A[l_1] [l_2]... [l_n]$ be a Traditional Multidimensional Array (TMA(n)) of size $[l_1, l_2,...,l_n]$ and $\langle x_1, x_2, ..., x_n \rangle$ be the subscripts of the array A; where $l_1, l_2,...,l_n$ is the length of dimension $d_1, d_2,..., d_n$ and $x_i = 0, 1..., (l_i - 1) (1 \le i \le n)$. We convert TMA(n) into a G2A $A'[l_1'][l_2']$ of size $[l_1', l_2']$ and subscripts $\langle x_1', x_2' \rangle$ where l_1' and l_2' are the length of dimension d_1' and d_2'; $x_1' = 0, ..., (l_1' - 1)$ and $x_2' = 0, ..., (l_2' - 1)$.

2.1 G2A for TMA(4)

Let $A[l_1][l_2][l_3][l_4]$ be a TMA(4) of size $[l_1, l_2, l_3, l_4]$. The location of the tuple $\langle x_1, x_2, x_3, x_4 \rangle$ can be linearized by $f(x_1, x_2, x_3, x_4) = x_1 l_2 l_3 l_4 + x_2 l_3 l_4 + x_3 l_4 + x_4$. In G2A, the 4 dimensions are converted into 2 dimensions. Figure 1 shows the G2A A $'[l_1'][l_2']$ for a TMA(4) A [2][3][3][2] where $l_1' = l_1 \times l_3$ and $l_2' = l_2 \times l_4$; $x_1' = x_1 l_3 + x_3$ and $x_2' = x_2 l_4 + x_4$; $x_i = 0, 1, ..., (l_i - 1) (1 \le i \le 4)$. Consider an element of A [1][1][2][0]. The corresponding element is therefore $A'[x_1'][x_2']$ (see Fig. 1) where $x_1' = 1 \times l_3 + 2 = 5$ and $x_2' = 1 \times l_4 + 0 = 2$. If an element in G2A is

Fig. 1 G2A realization from TMA(4)

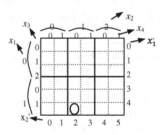

$A'[x_1'][x_2']$ is known then it's equivalent $A[x_1][x_2][x_3][x_4]$ can be found as $x_3 = x_1'$ % l_3 and $x_1 = x_1'/l_3$; $x_4 = x_2'$ % l_2 and $x_2 = x_2'/l_2$.

2.2   G2A for TMA(n)

The generalization of the TMA(n) $A[l_1]...$ $[l_n]$ to G2A $A'[l_1'][l_2']$ is as follows:

$$l_2' = \begin{cases} l_1 \times l_3 \times \cdots \times l_{n-1}, & \text{if } n \text{ is even} \\ l_1 \times l_3 \times \cdots \times l_n, & \text{if } n \text{ is odd} \end{cases}$$

$$l_2' = \begin{cases} l_2 \times l_4 \times \cdots \times l_n, & \text{if } n \text{ is even} \\ l_2 \times l_4 \times \cdots \times l_{n-1}, & \text{if } n \text{ is odd} \end{cases}$$

An element $A[x_1][x_2]...[x_n]$ in TMA(n) is equivalent to $A'[x_1'][x_2']$ where

$$x_1' = \begin{cases} x_1 l_3 l_5...l_{n-3}l_{n-1} + x_3 l_5 l_7...l_{n-3}l_{n-1} + \cdots + x_{n-3}l_{n-1} + x_{n-1}, & \text{if } n \text{ is even} \\ x_1 l_3 l_5...l_{n-2}l_n + x_3 l_5 l_7...l_{n-2}l_n + \cdots + x_{n-2}l_n + x_n, & \text{if } n \text{ is odd} \end{cases}$$

$$x_2' = \begin{cases} x_2 l_4 l_6...l_{n-3}l_{n-1} + x_4 l_6 l_8...l_{n-3}l_{n-1} + \cdots + x_{n-3}l_{n-1} + x_{n-1}, & \text{if } n \text{ is odd} \\ x_2 l_4 l_6...l_{n-2}l_n + x_4 l_6 l_8...l_{n-2}l_n + \cdots + x_{n-2}l_n + x_n, & \text{if } n \text{ is even} \end{cases}$$

For backward mapping, if an element in $A'[x_1'][x_2']$ is known then it's equivalent $A[x_1][x_2][x_3]...$ $[x_n]$ is as follows:

<u>n is even</u>

$x_n = x_2' \% l_n$

$x_i = ((...(x_2'/l_n).../l_{i+2})\% l_i, \quad i = 4,6,...,n-2$

$x_2 = ((...(x_2'/l_n)/l_{n-2}...)/l_6)/l_4$

$x_{n-1} = x_1' \% l_{n-1}$

$x_j = ((...(x_1'/l_{n-1}).../l_{j+2})\% l_j,$

$\qquad\qquad\qquad j = 3,5,...,n-3$

$x_1 = ((...(x_1'/l_{n-1})/l_{n-3}...)/l_5)/l_3$

<u>n is odd</u>

$x_n = x_1' \% l_n$

$x_i = ((...(x_1'/l_n).../l_{i+2})\% l_i, \quad i = 3,5,..n-2$

$x_1 = ((...(x_1'/l_n)/l_{n-2}...)/l_5)/l_3$

$x_{n-1} = x_2' \% l_{n-1}$

$x_j = ((...(x_2'/l_{n-1}).../l_{j+2})\% l_j,$

$\qquad\qquad\qquad j = 4,6,...,n-3$

$x_2 = ((...(x_2'/l_{n-1})/l_{n-3}...)/l_6)/l_4$

3   Comparison Between TMA and G2A for Matrix Operations

In G2A, the array cells are organized into chunks according to the number of dimensions. For a TMA(n) $A[l_1][l_2] ... [l_n]$ and its equivalent G2A $A'[l_1'][l_2']$ has the chunk size $|l_n \times l_{n-1}|$ and there are such $|l_1 \times l_2 \times \cdots \times l_{n-2}|$ chunks exists. Each chunk is a two dimensional array of size $[l_n, l_{n-1}]$. Figure 2 shows the data layout

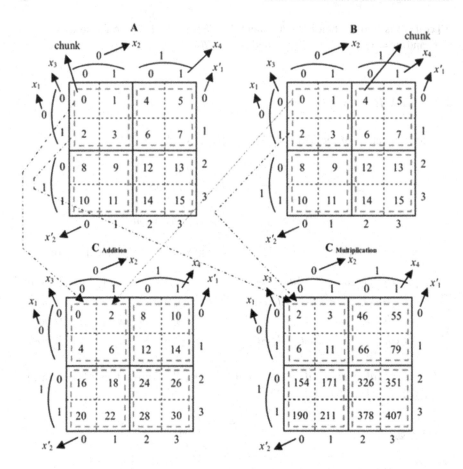

Fig. 2 Matrix-matrix addition and multiplication

separated into chunks for matrix- matrix addition/subtraction and multiplication for 4 dimensional matrices.

3.1 Matrix-Matrix Addition/Subtraction Algorithms

Let A and B be two TMA(3) of size $[l_1, l_2, l_3]$. The algorithm for matrix-matrix addition/subtraction $C = A + B$ based on the TMA(3) calculates the elements d_3 (i.e. x_3) fixing the elements of d_1 and d_2. Algorithm 1 and 2 show the matrix-matrix addition for TMA(n) and equivalent G2A respectively.

Algorithm 1:
matrix-matrix_addition_TMA_n
begin
   for $x_1 = 0$ to $(l_1$-1) do
     for $x_2 = 0$ to $(l_2$-1) do
       ....
   for $x_n = 0$ to $(l_n$-1) do
$C[x_1][x_2]...[x_n] = A[x_1]...[x_n] + B[x_1]...[x_n]$;
  End.

Algorithm 2:
matrix-matrix_addition_G2A
begin
   for $x'_1 = 0$ to $(l'_1$-1) do
     for $x'_2 = 0$ to $(l'_2$-1) do
     $C'[x'_1][x'_2] = A'[x'_1][x'_2] + B'[x'_1][x'_2]$;
End.

Let A' and B' be the equivalent G2A of a TMA(n). Since the G2A is a two dimensional representation of TMA(n), hence the matrix-matrix addition algorithm can be expressed as shown in Algorithm 2.

3.2 Matrix-Matrix Multiplication Algorithms

Let A and B be two TMA(n) of size $[l_1, l_2...l_{n-2}, l, l]$ (Since for matrix multiplication $l_{n-1} = l_n$). The algorithm for matrix-matrix multiplication $C = A \times B$ based on the TMA(n) is shown in Algorithm 3. Algorithm 4 shows the multiplication algorithm for G2A. The algorithm is row major order for G2A and each of the rows of C' is sequential. This row major ordering gives the facility that it decreases the access numbers of different elements of B' and the cache miss rate becomes lower.

Algorithm 3:
matrix-matrix_multiplication_TMA_n
begin
   for $x_1 = 0$ to $(l_1$-1) do
    for $x_2 = 0$ to $(l_2$-1) do
     for $x_3 = 0$ to $(l_3$-1) do
       ....
    for $x_{n-1} = 0$ to $(l$-1) do
     for $x_n = 0$ to $(l$-1) do
      for $i = 0$ to $(l$-1) do
$C[x_1][x_2]...[x_{n-1}][x_n] = C[x_1][x_2]...[x_{n-1}][x_n]$
  $+ A[x_1][x_2]...[x_{n-1}][i] \times B[x_1][x_2]...[i][x_n]$;
End.

Algorithm 4:
matrix-matrix_multiplication_G2A
begin
   for $x'_1 = 0$ to $(l'_1$-1) do
   begin
    m= $x'_1 - x'_1 \% l$
    for $x'_2 = 0$ to $(l'_2$-1) do
    begin
      n = $x'_2 - x'_2 \% l$
    for $i = 0$ to $(l$-1) do
    $C'[x'_1][x'_2] = C'[x'_1][x'_2] + A'[x'_1][n+i]$
  $\times B'[m+i][x'_2]$;
   end
End.

4 Experimental Results

The performances of our proposed algorithms are evaluated in Figs. 3 and 4 which show the execution time of matrix-matrix addition for TMA and G2A respectfully. The execution time is less for our proposed G2A scheme than TMA. This is because the algorithm for TMA has many loops than G2A based algorithm. Hence TMA based algorithm has higher cache miss rate than that of G2A based algorithm. The cache miss has direct influence to the performance because the processor needs to wait for the next data to be fetched from the next cache level or from the main memory. On the other hand G2A based algorithm improves the data locality that minimizes the cache miss rates as demonstrated in section [1–3].

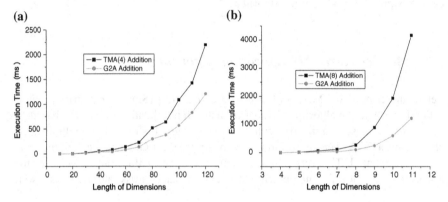

Fig. 3 Experimental performance for matrix-matrix addition for several dimensions; **a** Addition TMA(4) and G2A; **b** Addition TMA(8) and G2A

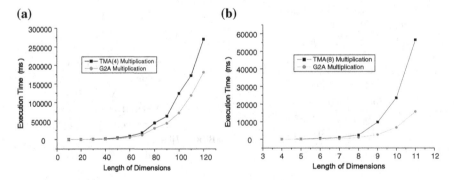

Fig. 4 Experimental performance for matrix-matrix multiplication for several dimensions; **a** Multiplication TMA(4) and G2A; **b** Multiplication TMA(6) and G2A

5 Related Works

Many techniques have been proposed in the literature for improving array computation. [1] proposed a scheme namely EKMR that represent a higher dimensional array into 2-dimensions (till 4-dimensional array only). The generalization of EKMR to n dimensions is a hierarchical structure that contains array of pointers. A technique based on loop transformation to improve the data locality for multi dimensional arrays is proposed in [2, 3] and demonstrated the usefulness for array operations. In [4] chunking, reordering, redundancy and partitioning of the large array are proposed to make efficient access on secondary and tertiary memory devices. Caching by chunk by chunk for improving performance is proposed by [5–7]. All the chunks are n dimensional with smaller length than the original array. Soroush and Balazinska [8] shows a technique for storing and analyzing multidimensional array by chunking but there is no generalization from higher dimensions.

6 Conclusion

In this paper, we have presented a new representation scheme for multidimensional arrays namely G2A. The G2A converts an n dimensional array into a 2 dimensional array. We designed efficient algorithm for matrix-matrix addition and multiplication to evaluate the proposed array representation. We found better results for G2A based algorithms than TMA based algorithms. G2A performs well because computation time for array indices is less as well as fewer cache miss rate in G2A. As a future work, we hope to develop a scheme for compressing multidimensional sparse arrays based on G2A.

References

1. Lin C-Y, Liu J-S, Chung Y-C (2002) Efficient representation scheme for multidimensional array operations. IEEE Trans Comput 51(3):327–345
2. Carr S, McKinley KS, Tseng C-W (1994) Compiler optimizations for improving data locality. In: Proceedings of the sixth international conference on architectural support for programming languages and operating systems, p. 252–262
3. McKinley KS, Carr S, Tseng C-W (1996) Improving data locality with loop transformations. ACM Trans Program Lang Syst (TOPLAS) 18(4):424–453
4. Sarawagi S, Stonebraker M (1994) Efficient organization of large multidimensional arrays In: Proceedings of 10th international conference on data engineering (ICDE), pp 328–386. Houston, Texas
5. Zhao Y, Deshpande P, Naughton JF (1997) An array-based algorithm for simultaneous multidimensional aggregates. In: Proceedings of SIGMOD Conference, pp 159–170
6. Deshpande P, Ramasamy K, Shukla A, Naughton JF (1998) Caching multidimensional queries using chunks. In: Proceedings of the ACM SIGMOD conference on management of data, pp 259–270

7. Steinbach M, Ertöz L, Kumar V (2004) The challenges of clustering high dimensional data. New directions in statistical physics, pp 273–309. Springer, Berlin
8. Soroush E, Balazinska M (2011) ArrayStore: a storage manager for complex parallel array processing. In: Proceedings of ACM SIGMOD international conference on management of data, pp 253–264

A Portable and Platform Independent File System for Large Scale Peer-to-Peer Systems and Distributed Applications

Andreas Barbian, Stefan Nothaas, Timm J. Filler and Michael Schoettner

Abstract Virtual microscopy is an evolving medical application used for teaching and learning at universities. We have developed a peer-to-peer based solution called Omentum, aiming at bringing virtual microscopy to an Internet-scale community. Omentum has to manage more than 10,000 large proprietary microscopic images that are converted to easily dividable JPEG-trees, each consisting of millions of very small-scaled image parts. In this paper we propose a portable and platform independent user space file system (USPFS) for addressing the application-specific access patterns, security concerns, and data integrity. USPFS is able to efficiently manage huge capacities (roughly 9×10^{18} slices with 9,000 Petabytes each) with a theoretically infinite number of storable objects while providing highly important platform independency, data integrity checks as well as an easily extendable API. The evident metadata overhead is only 0.3 % and the performance evaluation shows promising results for both read and write operations.

Keywords File systems · Peer-to-peer systems · Distributed applications

1 Introduction

According to Rojo et al. a virtual microscope is a concept that "includes different aspects [...] spanning from image acquisition to visualization systems" [10]. In the past decade different use cases with a CD-ROM-based image distribution emerged [2], nowadays a client-server architecture [5] is most commonly seen. Usually, the digitalized slides used in a virtual microscope are several gigabytes in size, but to

A. Barbian (✉) · T.J. Filler
Department of Anatomy, Heinrich-Heine-University, Duesseldorf, Germany
e-mail: andreas.barbian@hhu.de

A. Barbian · S. Nothaas · M. Schoettner
Department of Computer Science, Heinrich-Heine-University, Duesseldorf, Germany

© Springer Science+Business Media Singapore 2016   47
J.J.(Jong Hyuk) Park et al. (eds.), *Advances in Parallel and Distributed Computing and Ubiquitous Services*, Lecture Notes in Electrical Engineering 368,
DOI 10.1007/978-981-10-0068-3_6

constitute as a fully equivalent substitute for a traditional histology class, several hundreds or ideally thousands of images are needed.

The distributed virtual microscope Omentum is our approach to apply current peer-to-peer research results to this application domain [6]. In its design the virtual slides have to be distributed, which leads to many peer-to-peer participants providing different image parts. In general, these image parts consist of different files stored somewhere on a peers' local hard drives.

Without knowledge of the underlying operating system on these devices, it is nearly impossible to rely on data security features provided by the operating system.

Furthermore, it cannot be ensured that every peer uses the same file system. Hence, basic data integrity checks cannot be used because they may simply be unavailable on some file systems. However, data integrity checks are essential, because an intentional or even accidental data corruption needs to be detected due to replication algorithms that may otherwise override correct data on other peers, too.

We address these challenges by presenting a novel platform-independent and adaptable file system that is capable of handling a tremendous amount of storing space (roughly 9×10^{18} slices with 9 K Petabytes each) and theoretically infinite storable objects.

2   User Space File System

The file system itself needs to be supported by many desktop operating systems. The User Space File System (USPFS) is an abstract design for a hierarchical file system and was developed while focusing on data integrity, portability and the need to handle a huge number of storable files.

A file system usually uses a designated partition or volume, exclusively reserved on some kind of physical storage container for its organization. This leads to a simple physical limitation, as a file system cannot occupy more blocks than provided by its container. Some file systems rely on a volume manager to combine different containers and present it as a single addressable space to the accessing operating system [3]. Likewise USPFS should not be limited and hereby requires a distinct managing level for supporting multiple data storage containers of various types.

To ensure portability to most desktop platforms, USPFS's data is stored in files on any available file system hosted by the operating system—like HDFS [11].

As storing many small files is necessary in Omentum in the first place, USPFS' main goal is to support a large number of stored objects. Therefore, within USPFS, addressing data is handled in two steps. At first, a position in the volume is described by its slice index and then by its byte position in the selected slice. Each of these two indices is stored in a 64-bit wide variable. Although this limits the capacity of a single slice to 2^{63} bytes (roughly 9 K petabytes) addressable space including metadata, the possibilities with 2^{63} (roughly 9×10^{18}) slices are sufficient for Omentum and surely for many other applications.

In this context, the number of files in a single directory and in the entire file system has to be considered: In Omentum up to 0.3 million files per virtual slide have to be handled (see Fig. 1a) with small images being in the focus of interest (see Fig. 1b). Even with only a few thousand slides, it implies the necessity of storing several billion files. As USPFS does not make use of fixed index structures, there is practically no limit to the number of objects stored in the file system.

A single instance of USPFS consists of a *superblock*, a *memory manager* and a number of *slices* stored as separate files. These files form the complete USPFS *volume*. By splitting up the volume's data into slices, the user can store them on different partitions or physical hard drives, thus not being limited to the boundaries or capacity of a single host file system.

The *superblock* identifies the file system instance as well as the actual volumes storing the data. It stores the USPFS version used for this instance, the name of the volume, the block size, a pointer to the root *Inode* as well as information about the *memory manager* and *filters* chosen on creation. The volume information covers the slice index and the maximum size of each slice.

A separate *memory manager* takes care of managing free and allocated blocks of memory and hereby replaces fixed structures within the volume for block occupation. This removes limits introduced by other file systems like the EXT family [4], which reserve a partition of memory at the beginning of the volume's address space for this purpose. The manager writes its data to a separate file, which is not part of the slices. Depending on the specific use case, it is possible to implement a highly customized memory manager and choose it upon volume creation. This can be very useful to optimize the performance of the file system and the space the memory manager uses.

An IO subsystem, which can connect to different underlying sources to read from and/or targets to write to, provides access to the volume. The default source is a normal file hosted on a file system the user can access via the host operating

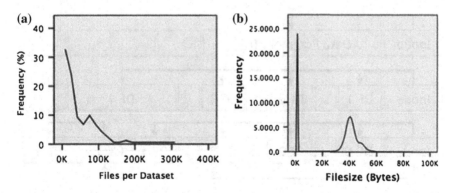

Fig. 1 Distribution of files and their sizes in Omentum's datasets. **a** A small database, **b** a single dataset

system. By providing a customizable API, the user can extend this feature with his own implementations like accessing data via a network socket. Furthermore, the data stream that is read from the source and written back to it, can be piped through filters, which can be activated upon volume creation, like e.g. known from FUSE [12].

Here, a *filter* describes a set of two functions, which modify the data stream used for reading and writing. Multiple filters can be concatenated for altering data. These filters may be activated upon creation of a new volume. They typically provide functions like data compression or encryption depending on their implementation. In order to provide better consistency checks for stored data, the user can enable a checksum algorithm upon volume creation. The calculated checksum is stored in the file's metadata and is verified on file access.

3 Metadata Structures

All data is organized in blocks of the size described within the superblock (a multiple of 512 bytes per block), depending on the used memory manager. Thus, the memory management is simplified, streamlined and made more efficient.

The *Inode* indexes each file or every part of it and stores any information regarding a single file. It provides the file type (normal file, directory, symbolic link), user-defined flags, last access and modification date, any access flags, ID for owner and group, file name, total size and checksum of the file, like POSIX. It also stores direct pointers for up to four fragments for fast access (see Fig. 2), like extents in EXT4 [7].

A *fragment* describes one block of memory in the file system with a length of an integral multiple of the set block size. With the actual memory layout of the file system, the data can be stored in only one fragment during a single write operation regardless of its length. This enables consecutive reads and writes on it, which

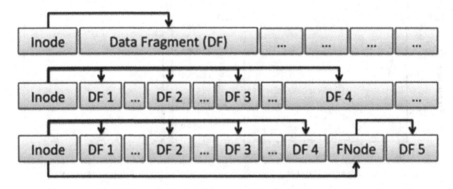

Fig. 2 Simplified example for three possible layouts of metadata and data of a single file within a slice

increases the speed of these operations. Given a bigger file's fragmentation, another pointer in the Inode structure references another Fnode.

Each *Fnode* further stores pointers to additional fragments making them indirect references. Multiple Fnodes are double linked, forming a chain of Fnodes with the first one being referred to by the Inode. Organized like this, the maximum numbers of fragments for a single file is theoretically unlimited. This results in a maximum file size, including the file's metadata, of nearly the total size of the volume.

4 Implementation

Java was chosen as the programming language for the implementation to match the key point of portability avoiding to re-compile the code on every platform. Additionally, regarding the Omentum project that is likewise written in Java, this eased the process of integration. Nevertheless, it is possible to use any other language to implement the specification on other platforms.

The implementation uses only classes from the native Java packages *java.io*, *java.nio* and *java.util* and does not have any other external dependencies. The organisation and interaction between the key components is visualized in Fig. 3 and will be referred to by selected components.

The interface *IODataStream* has to be implemented, depending on the source to read/write data from/to, to provide access to any kind of data. The implementation *IODataStreamFile* utilizes Java's *RandomAccessFile* for plain file access. *IODataStream* can be implemented in a different fashion to allow reading from different sources like a network socket or other non file like locations.

Every metadata structure (superblock, Inode, Fnode) is represented by its own class. These classes need to be serialized and de-serialized to read them from or write them back to an *IODataStream* source.

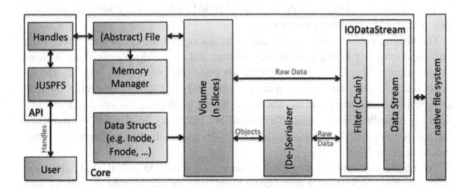

Fig. 3 Organization and interaction of the key components of the JUSPFS implementation

The (abstract) *File*, as seen in Fig. 3, is part of the core of JUSPFS. Most of the file system's logic is handled in that class. This includes managing metadata for a single file, interaction with the memory manager for freeing and allocating memory related to any operations affecting memory, reading existing metadata/data from and writing back newly created metadata and data back to the volume. The File class takes care of automatically extending metadata if the file is forced to be split into fragments due to memory layout on the volume. The result is a generic interface to write any kind of file type to the volume.

The class *FileHandle* is the actual implementation of a normal binary file for JUSPFS, though it isn't too different from the abstract File. JUSPFS is designed as a hierarchical file system, thus it needs an implementation of a directory. *DirectoryHandle* implements a directory utilizing a hash map for storing its entries. Using the standard hash map implementation of Java allows storing 2^{31} (roughly 2 billion) entries in each directory.

The file system is easily extendable and a user can implement custom handles for any required use case.

5 Evaluation

Especially for file systems, it is important to evaluate not only the speed of reading and writing, which will be presented later in this section, but also the efficiency for storing the data regarding the ratio of metadata to payload.

A typical dataset from Omentum is used to evaluate the performance of JUSPFS. The metadata size of the entries is not an issue if seen in relation to the amount of files being stored in a folder. The dataset contains 98,170 files and has a total payload of 3,111,791,334 bytes. The distribution of files within the dataset is equivalent to Fig. 1a with a minimum of 289 bytes, an average of 31,698 bytes and maximum of 86,751 bytes.

The total amount of metadata necessary is around 7 MB only. The information from the memory manager for the volume is added with around 4 MB resulting in a total metadata-to-payload-ratio of 0.3 %.

The writing performance was measured to ensure that the speed does not change with the increasing filling state of the volume. All tests were executed on a clean system running the OS and our application only. Despite of this, Fig. 4a shows some nearly equidistant peaks that are obviously related to operating system specific task scheduling during the tests. These peaks can be observed on any platform regardless of the used file system but may vary in distance. The overall average writing speed per byte in JUSPFS is 16 ns resulting in an average writing performance of 59.60 MB/s.

To ensure constant reading performance, we executed a series of tests that randomly read files from a previously filled file system at various lengths and positions in the file system. A worst-case scenario where we do not profit from read caching at all, would result in an average reading speed of 15.41 MB/s. Although,

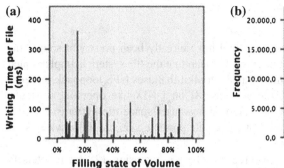 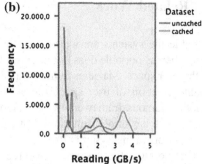

Fig. 4 Performance evaluation for writing and reading files from JUSPFS regarding the volume's filling state and caching. **a** Writing performance for many files regarding the filling state of a JUSPFS volume, **b** reading performance for a series of random reads with and without caching

this may seem very low, the required speed for a single node to serve 20 hosts requesting data is around 6 MB/s. This value has been gained from ongoing tests throughout more than a year. The replication strategy tries to level incoming requests and in case too much traffic is created on a single node, the data is replicated once more to distribute the load. Using typical data-request patterns, which involves read caching due to multiple requests of the same file, the average reading speed is 1 ns, leaving an average reading throughput of 953 MB/s (see Fig. 4b).

Caching has a strong influence on the result of the reading and writing tests and this leads to the question of comparability of results under various operating system and file system combinations. We executed a number of tests using three different operating systems with OS-specific file systems. As these tests did not show platform-specific Java differences, we executed another one that uses the same file system (FAT32) on different operating systems. The results, normalized by the highest performance, are presented in Table 1. Optimal performance was achieved using Linux with EXT4, whereas the Windows and NTFS combination performed worst. The differences in performance can be explained by the fact that the operating systems and file systems natively support a different number of security features (e.g. user access control, group flags and logging), which are activated by default. For comparison, we did not take any measures to optimize the performance of any used file system.

Table 1 Normalized performance results of OS and file system combinations hosting JUSPS, with 1.0 indicating best performance

Operating system	OS file system	OS FS	FAT 32
Ubuntu 13.04	ext4	1.00	0.97
MacOS 10.7.5	MacOS Extended (Journaled)	0.88	0.84
Microsoft Windows 7	NTFS	0.54	0.76

6 Related Work

Portable file systems are well known and have already been presented some time ago, whereas portable does not necessarily mean that the file system is implemented in the user space. Mazières demonstrated a toolkit that uses NFS loopback server to enable creation of user space file systems [8] on UNIX-like operating systems. Another famous framework for developing new user space file systems is FUSE [12], which has implementations that e.g. create portable version of file systems for different operating systems [9].

Our portable file system is inspired by ZFS [3] for using 128 bit pointers to address data and achieve a very high capacity. Additionally, various versions of the EXT file system [7] had influence on our design, as we tried to overcome the 16 GB limit for a single file (using a 4 K block size) and its maximum capacity of 10^{18} bytes. Furthermore, the extents used in the Journaled File System (JFS) [1] inspired a performance increase for reading and writing while avoiding a limited number of storable files.

The Hadoop Distributed File System [11] is written in Java, too. It focuses on the distributed storage of mainly very large files, which does not match our application case with millions of small files.

7 Conclusion

We have presented the design and implementation of the novel user-level file system USPFS (implemented in Java) addressing the specific requirements of the application domain of virtual microscopy, namely: portability, security, data integrity, storage capacity (huge number of small files per directory and per volume). The file system has an open architecture allowing to easily addressing emerging needs of new application domains. The experiments and evaluations show that USPFS works well on different desktop operating systems with different underlying native file systems. Furthermore, USPFS has a very low meta-data overhead (around 0.3 %). The read and write throughput is fast and not impacted by the additional layer of USPFS.

References

1. Best S (2000) Jfs log: how the journaled file system performs logging. In: Proceedings of the 4th annual Linux showcase and conference, pp 163–168, October 2000
2. Blake CA, Lavoie HA, Millette CF (2003) Teaching medical histology at the university of south Carolina school of medicine: transition to virtual slides and virtual microscopes. Anatom Rec (New Anat) 275B:196–206

3. Bonwick J, Ahrens M, Henson V, Maybee M, Shellenbaum M (2003) The zettabyte file system. Technical report, SUN Microsystems

4. Card R, Ts'o T, Tweedie S (1994) Design and implementation of the second extended filesystem. In: Brokken FB, Kubat K (eds) Proceedings of the first Dutch international symposium on Linux. State University of Groningen, December 1994

5. Heidger PM Jr, Dee F, Leaven T, Duncan J, Kreiter C (2002) Integrated approach to teaching and testing in histology with real and virtual imaging. Anatom Rec (New Anat) 269(2): 107–112

6. Jaegermann A, Filler TJ, Schoettner M (2013) Distributed architecture for a peer-to-peer-based virtual microscope. In: Dowling J, Taiani F (eds) Distributed applications and interoperable systems. Springer, pp 199–204, June 2013

7. Mathur A, Cao M, Bhattacharya S, Dilger A, Tomas A, Vivier L (2007) The new ext4 filesystem: current status and future plans. In: Proceedings of the Linux symposium, vol 2, pp 21–33, Ottawa, Ontario, Canada, June 2007

8. Mazières D (2001) A toolkit for user-level file systems. In: Proceedings of the 2001 USENIX annual technical conference, Boston, Massachusetts, USA, June 2001

9. Rajgarhia A, Gehani A (2010) Performance and extension of user space file systems. In: Proceedings of the 2010 ACM symposium on applied computing, Sierre, Switzerland, March 2010. ACM New York, NY, USA, pp 206–213

10. Rojo MG, García GB, Mateos CP, García JG, Vincente MC (2006) Critical comparison of 31 commercially available digital slide systems in pathology. Int J Surg Pathol 14(4):285–305

11. Shvachko K, Kuang H, Radia S, Chansler R (2010) The hadoop distributed file system. In: Proceedings of the IEEE 26th symposium on mass storage systems and technologies (MSST). IEEE, pp 1–10, May 2010

12. Szeredi M File systems in userspace. http://fuse.sourceforge.net

OCLS: A Simplified High-Level Abstraction Based Framework for Heterogeneous Systems

Shusen Wu, Xiaoshe Dong, Heng Chen and Bochao Dang

Abstract In contrast with the increasing popularity of heterogeneous systems, programming on these systems remains complex and time-consuming. Developers have to access heterogeneous processors through explicitly and error-prone operations provided by low-level approaches like OpenCL. We present OCLS (OpenCL Simplified), a high-level abstraction based framework and its implementation as a minimal library on the top of OpenCL. OCLS shields hardware details, simplifies the development process and handles the environment configuration and data movement implicitly. Its APIs act like ordinary functions and require little prior training. OCLS thus reduces heterogeneous programming effort and relieves the programmers of low-level programming. We evaluated OCLS across a set of different benchmarks. The size of benchmarks rewritten in OCLS reduced by an average ratio of 35.4 %. In the experiment on both GPU and Intel MIC platforms with data sets in different size, OCLS yielded better performance than original OpenCL programs and showed a good stability and portability.

Keywords Heterogeneous programming · OpenCL · OCLS · Abstraction

1 Introduction

Heterogeneous systems employing different kinds of accelerators/co-processors are continuously dominating the high performance computing area according to the Top500 list [1]. With a 33.86PFlop/s peak performance, the Tianhe-2 supercomputer which uses Intel Xeon CPUs and Xeon Phi co-processors is currently the

S. Wu (✉) · X. Dong · H. Chen · B. Dang
School of the Electronic and Information Engineering, Xi'an Jiaotong University,
Xi'an, Shaanxi, P. R. China
e-mail: wuss153@stu.xjtu.edu.cn

© Springer Science+Business Media Singapore 2016   57
J.J.(Jong Hyuk) Park et al. (eds.), *Advances in Parallel and Distributed Computing and Ubiquitous Services*, Lecture Notes in Electrical Engineering 368,
DOI 10.1007/978-981-10-0068-3_7

fastest system worldwide. The cost-effectiveness, power-efficient and high performance accelerators/co-processors have caused the shift from homogeneous programming to heterogeneous programming. However, the state-of-art heterogeneous programming methods [2, 3] could be disappointing.

The current de facto standard for heterogeneous computing is OpenCL [4]. With vendor provided runtime support, OpenCL is available on a wide range of architectures. The theme of OpenCL programming is to offload parallel computations (kernels) on devices. To archive that, applications have to start with device query and environment configuration. Resource allocation on device, data transfer from host to device and vice verse after the kernel execution are also needed. All the operations above including kernel launch are done explicitly through the OpenCL APIs. Programmers deal with hardware details which may limit the portability of the program. This leads to tedious and error-prone code and makes heterogeneous programming complex and difficult.

We present a high-level abstraction based framework called OCLS to simplify the programming on heterogeneous systems. OCLS framework provides a single virtual processor abstraction for all heterogeneous systems and hides the hardware details for programmers through a library. The OCLS library encapsulates the OpenCL APIs and realizes automatic environment configuration. Data movements are handled at runtime implicitly with the help of a kernel data type defined in OCLS. The OCLS framework enables programmers to carry out heterogeneous computing in an ordinary program with little extra effort.

2 Related Works

OpenCL provides rich low-level APIs for heterogeneous programming. It builds a solid foundation for high-level extensions. OpenACC [5] is a high-level directive based approach currently targets at single device. It reduces the difficulty of programming at the expense of performance and flexibility.

JSeriesCL [6] defines ParameterGPU class to simply the OpenCL execution and using associated data and thread attributes to decide the size of work-groups and work-items. Using it requires heavy prior training and the code has poor readability.

The SOCL [7] framework provides a unified OpenCL platform for multi-device system. It ease the restrictions on platforms, contexts and command queues of OpenCL without changing the development process.

The Skeleton computing language (SkelCL) [8] is a high-level extension of OpenCL for multi-GPU system. It introduces parallel container data types and parallel skeletons to realize automatic data distribution and parallel computation. Although it's powerful, it has a limited scope of applications. It's learning cost could be high.

The VirtCL [9] framework provides a single high-level abstraction for multiple devices. It implements a front-end library instead of the OpenCL APIs. Since the VirtCL is presented to solve the problems of memory inconsistency and device contention, its library only partly conceals the OpenCL programming details.

3 OCLS Framework

3.1 OCLS Abstraction Layer

The initial motivation of OCLS is to reduce the difficulties to exploit the massively parallel computing capability of heterogeneous system. Application developers demand a high-level abstraction as they are suffering from low-level programming. OCLS provides a unified abstraction layer between programmers and heterogeneous systems and implements such abstraction on the top of OpenCL as illustrated in Fig. 1. For programmers, they are interacting with a virtual processor through OCLS library. There is little difference from programming on a multi-core processor except that the parallel scale can be several orders of magnitude higher. They will concentrate on designing and implementing parallel algorithms into kernels and call the OCLS APIs to execute them. The execution is controlled by the parameters passed to the API function. All the details of low-level programming on various heterogeneous processors are handled implicitly. With C compatible library and OpenCL kernel language, OCLS integrates heterogeneous programming into ordinary programs.

Fig. 1 OCLS abstraction layer

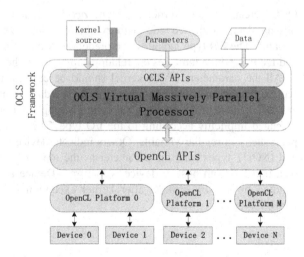

3.2   OCLS Library

The OCLS library consists of four primary functions and three assistant functions. It encapsulates OpenCL APIs and minimizes the development process of OpenCL. The four primary functions indicates the procedures in OCLS programming: initialization, kernel execution, execution finalization and termination.

The *ocl_Init()* function queries all available platforms and devices and selects a best device to initialize the environment configuration. It then compile the kernel source for execution. If the kernel source is stored in kernel files, call the assistant function *prog_Src()* to read them in ahead of *ocl_Init()*. It takes the kernel compile option as parameter and just need to be called once.

Programmers use the *ocl_Runkernel()* function to launch a kernel. They just give the name of the kernel and specify the execution scale (NDRange in OpenCL) along with the kernel parameters. *ocl_Runkernel()* then creates the kernel, set the kernel arguments and run it on the selected device.

The *ocl_Finkernel()* function finalizes the kernel execution. It sets a synchronization point and automatically handles the data movement. It should be called before the subsequent calculations using the results and need not to be paired with *ocl_Runkernel()*.

The *ocl_End()* function terminates the OCLS programming by releasing allocated memory space on both the host and the device. The shared OpenCL objects like context, program and kernel created earlier are also purged.

3.3   Kernel Data Type and Data Movement

OCLS creates a new kernel data type *ocl_kdata*. It consists of a pointer to the original data, the device side memory object, a data size variable and the I/O type variable. The I/O type includes five predefined value: NORMAL, TEMP, IN, OUT and INOUT which indicates the relationship between the kernel execution and data.

All the parameters of the kernel should be created in *ocl_kdata* type and initialized using assistant function wrapper(). The data pointer, data size and I/O type are specified by programmer when calling wrapper(), the device memory object is managed implicitly according to the I/O type. The NORMAL type states that the parameter is a regular variable. Others indicate device side memory allocation. IN and INOUT types state the parameter as the input data to the kernel, a implicitly data transfer from host to device is incurred. Device to host data movements are handled by *ocl_Finkernel()* after kernel execution on the data with OUT and INOUT type.

Kernel execution often needs extra memory space for intermediate results. Those parameters are stated as TEMP type. OCLS also provide assistant function *flush_Data()* for user controlled random data transfer for sake of flexibility.

3.4   Runtime Data Structures

OCLS introduces several data structures at runtime to provide convenience and eliminate redundant operations.

The *parallel capacity* refers to the product of the compute units amount and clock frequency of a device. It's the criterion that OCLS uses to select the best device.

After the kernel execution, kernel data with OUT or INOUT type is automatically pushed into a shared *output stack*. A ocl_Finkernel() function called later empties the stack and copies the results back.

The same kernel will be created in every call to function *ocl_Runkernel()* If it is launched repetitively. It also affects the output stack. A *history pointer* is introduced to solve this problem by recording the last executed kernel. If the same kernel is launched, *ocl_Runkernel()* will skip the kernel creation and pushing operations.

A *buffer list* is used to indicate the location of each buffer in device memory. When a buffer is created in *wrapper()*, a pointer to the buffer is added to the tail of the buffer list. When calling *ocl_End()*, it frees allocated memory space according to the list.

4   Case Study

An example of implementing the same vector addition algorithm in both OCLS and OpenCL is presented to demonstrate the use of OCLS and its advantages. Algorithm 1 show the kernel source code of vector addition which is used for both Algorithm 2 and 3. The initialization of the three integer vectors A, B and C with a same size specified by a variable *datasize* is omitted.

Algorithm 2 and 3 both archive the function of executing vector addition on a device in parallel. But the original OpenCL program has to specify the device type explicitly which limits its portability and needs an extra 29 lines in source code to accomplish the same task. It omits all the error handling which is done implicitly in OCLS and the readfile() function used to read in kernel file is also undefined. The example shows that using OCLS can reduce the programming effort significantly with user-friendly APIs and gives a brief look at the portability and reliability of OCLS.

Algorithm 1: Vecadd kernel

1. __kernel void vecadd(global int *a,\
   global int *b,global int *c)
2. {
3.   int id = get_global_id(0);
4.   c[id] = a[id] + b[id] ;
5. }

Algorithm 2: Vecadd in OCLS

1. ocl_kdata da,db,dc;
2. prog_Src("kernel.cl");
3. ocl_Init(NULL);
4. da = wrapper(A,datasize,IN);
5. db = wrapper(B,datasize,IN);
6. dc = wrapper(C,datasize,OUT);
7. ocl_Runkernel("vecadd",1,&elements,NULL,3,da,db,dc);
8. ocl_Finkernel();
9. ocl_End();

Algorithm 3: Vecadd in OpenCL

1. cl_int err;
2. cl_platform_id platform;
3. cl_device_id device;
4. cl_device_type devicetype;
5. cl_context context;
6. cl_command_queue queue;
7. cl_program program;
8. cl_kernel kernel;
9. char * clProgStr = "";
10. cl_mem dA,dB,dC;
11. devicetype = CL_DEVICE_TYPE_GPU;
12. clGetPlatformIDs(1,&platform,NULL);
13. clGetDeviceIDs(platform,devicetype,1,&device,NULL);
14. cl_context_properties prop[] = {CL_CONTEXT_\
    PLATFORM, (cl_context_properties) platform, 0};
15. context = clCreateContext(prop,1,&device,NULL,NULL,\
    &err);
16. queue = clCreateCommandQueue(context,device,0,&err);
17. clProgStr = {readFile("kernel.cl")};
18. program = clCreateProgramWithSource(context,\
    1,(const char **) &clProgStr, &srcLength, &err);
19. clBuildProgram(program,1,&device,NULL,NULL,NULL);
20. dA = clCreateBuffer(clContext,CL_MEM_READ_ONLY\
    ,datasize,NULL,&clStatus);

21. dB = clCreateBuffer(clContext,CL_MEM_READ_ONLY\
    ,datasize,NULL,&clStatus);
22. dC = clCreateBuffer(clContext,CL_MEM_WRITE_\
    ONLY,datasize,NULL,&clStatus);
23. clEnqueueWriteBuffer(queue,dA,CL_TRUE,0,datasize,\
    &A,0,NULL,NULL);
24. clEnqueueWriteBuffer(queue,dB,CL_TRUE,0,datasize,\
    &B,0,NULL,NULL);
25. clEnqueueWriteBuffer(queue,dC,CL_TRUE,0,datasize,\
    &A,0,NULL,NULL);
26. clSetKernelArg(kernel,0,sizeof(cl_mem),(void*)&dA);
27. clSetKernelArg(kernel,0,sizeof(cl_mem),(void*)&dB);
28. clSetKernelArg(kernel,0,sizeof(cl_mem),(void*)&dC);
29. clEnqueueNDRangeKernel(queue,kernel,1,NULL,\
    &elements,NULL,0,NULL,NULL);
30. clEnqueueReadBuffer(queue,dC,CL_TRUE,0,datasize,\
    &C,0,NULL,NULL);
31. Free(clProgStr);
32. clReleaseKernel(kernel);
33. clReleaseProgram(program);
34. clReleaseMemObject(dA);
35. clReleaseMemObject(dB);
36. clReleaseMemObject(dC);
37. clReleaseCommandQueue(queue);
38. clReleaseContext(context);

---

5 Evaluation

All of the evaluations were conducted on both a GPU server with two Xeon E5520 CPUs, 12G RAM and four NVIDIA Tesla C1060 GPUs running CUDA 6.5 [10] and CentOS 6.5 with linux Linux 2.6.32-431 kernel and a Intel MIC server with two Xeon E5-2670 CPUs, 64G RAM and two Xeon Phi 7110P co-processors running Intel OpenCL runtime 14.2 with MPSS 3.3.4 and Red Hat Enterprise Linux Server 6.3 with linux 2.6.32-279 kernel. The OpenCL version on both platform is OpenCL 1.2. The benchmarks we used were collected from the Parboil benchmark suite [11]. The benchmarks and corresponding data sets are shown in Table 1.

Table 1 Benchmarks and data sets

Benchmark	Problem size (small)	Problem size (large)
bfs	270,926 nodes	1,000,000 nodes
cutcp	5943 atoms	–
histo	996 w × 1040 h 20 iterations	–
lbm	2.16×10^6 cells 100 iterations	–
mri-q	32768 pixels using 3072 samples	262144 pixels using 2048 samples
sgemm	Matrix size: 128 × 96 96 × 160	Matrix size: 1024 × 992 992 × 1056
stencil	Grid size: 128 × 128 × 32	Grid size: 512 × 512 × 64

5.1 Code Size Comparison

All the benchmarks were rewrote using OCLS without any changes or optimization to the algorithm. Figure 2a shows the source code size of the benchmark and OCLS programs. Figure 2b shows the normalized source code size comparison. The reduction in coda size is related to the program feature since OCLS only reduces the code of parallel relevant parts. Benchmark *cutcp* spends much effort on data processing and serial computing thus produces the lowest reduction ratio. OCLS archives an average 8.97 KB reduction in code size. The average reduction ratio is 35.4 %.

5.2 Performance

The OpenCL programs in Parboil benchmark suite target GPU. It needs some manual modification in environment configuration to get them run on MIC platform. However, benchmark *cutcp* kept failing on GPU due to the OpenCL implementation issues in CUDA, it was removed during the GPU test. Benchmark *histo* incurred hardware exception and segmentation fault during the MIC test, the program returned with incorrect result. However, the OCLS version of *histo* ran properly.

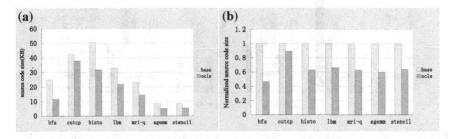

Fig. 2 Source code size comparison (*base* original OpenCL program)

Although OCLS introduces overhead in device querying and extra data structures creation, it eliminates redundant operations introduced by encapsulation with the runtime data structures and avoids unnecessary data transfer in original benchmark program. It yields better performance on both GPU and MIC as shown in Fig. 3. Benchmark *cutcp*, *histo*, *lbm*, *mri-q* used *small* data set. Benchmark *bfs*, *sgemm*, *stencil* used *large* data set.

5.3 Stability

We also evaluated the stability of OCLS programs using different size data sets provided by the benchmark suite. Figure 4 illustrates the test result. We can see that the performance of OCLS is always better than original OpenCL programs. The trend in the execution time changes of OCLS follow that of OpenCL. OCLS performed as stable as OpenCl with different problem size and kept its advantages. It's also very interesting to find the differences between GPU and MIC of their behavior with different data size and application.

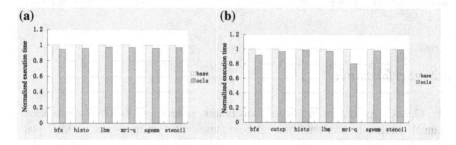

Fig. 3 Normalized execution time comparison

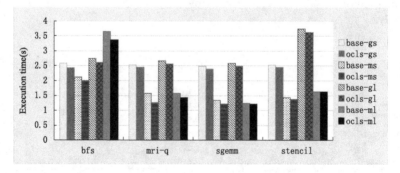

Fig. 4 Execution time with different data sets (-s: small, -l: large) on GPU and MIC (-g: GPU, -m: MIC)

6 Conclusion

We have proposed a simplified high-level programming framework called OCLS for heterogeneous system. It allows programmer to exploit the massively parallel computing capability of various heterogeneous processors in an ordinary program without concerning the hardware details. This is archived through the OCLS library which encapsulates the OpenCL APIs. The OCLS library minimizes the development process and realizes automatic environment configuration and data movement. The comparison with original OpenCL programs shows that using OCLS can reduce the amount of code and the programming effort significantly. In the experimental evaluation on GPU and MIC with different data size, OCLS showed a stable and better performance than the benchmarks. The experiments also demonstrated the portability and stability of OCLS.

Acknowledgments This work is supported by the National Natural Science Foundation of China (NSFC) under Grant No.61173039, and the National High Technology Research and Development Program (863 Program) of China under Grant No. 2012AA010904.

References

1. Top500.org. http://www.top500.org/
2. Javier Diaz, Camelia Munoz-Caro, Alfonso N (2012) A survey of parallel programming models and tools in the multi and many-core era. IEEE Trans Parallel Distrib Syst 23(8):1369–1386
3. Brodtkorb Andre R, Christopher Dyken, Hagen Trond R et al (2010) State-of-the-art in heterogeneous computing. Sci Program 18:1–33
4. The OpenCL specification. https://www.khronos.org/opencl/
5. OpenACC–directives for accelerators. http://www.openacc-standard.org/
6. de Souza Rosa Gomes R, Figueiredo JM, Martins CA et al (2014) A framework for automating the configuration of OpenCL. Environ Model Softw 53:81–86
7. Henry S, Denis A, Barthou D, Counilh M-C, Namyst R (2014) Toward OpenCL automatic multi-device support. In: Euro-Par 2014, LNCS, vol 8632. Springer, Heidelberg, pp 776–787
8. Steuwer M, Gorlatch S (2014) SkelCL: a high-level extension of OpenCL for multi-GPU systems. J Supercomput 69:25–33
9. You Y-P, Wu H-J, Tsai Y-N et al (2015) VirtCL: a framework for OpenCL device abstraction and management. In: 20th ACM SIGPLAN symposium on principles and practice of parallel programming. ACM, New York, pp 161–172
10. CUDA toolkit. https://developer.nvidia.com/cuda-toolkit
11. Parboil Benchmarks. http://impact.crhc.illinois.edu/Parboil/parboil.aspx

Hierarchical Caching Management for Software Defined Content Network Based on Node Value

Jing Liu, Lei Wang, Yuncan Zhang, Zhenfa Wang and Song Wang

Abstract Architecture combining Content-Centric Network (CCN) and Software-Defined Network (SDN) has gradually attracted more attention. We have realized a prototype of CCN using Protocol-Oblivious Forwarding (POF), called Software-Defined Content Network (SDCN). And SDCN does not rely on the IP. CCN supports the unique in-network caching, so that caching strategies become a challenge. CCN lacks of global recognition for the whole network that leads to unreasonable resource allocation. This paper focuses on collecting topology information in SDCN and constructing a hierarchical cache model of the network. We discuss how to distribute cache capacity based on node value under the total storage budget for the network. A cache strategy based on node value is also proposed, which places contents on the nodes with different values according to their popularity. Experimental results show that the cache performance in SDCN was improved.

Keywords POF · CCN · Software-defined content network · Hierarchical cache allocation · Node value

J. Liu · L. Wang (✉) · Y. Zhang · Z. Wang · S. Wang
Department of Automation, University of Science
and Technology of China, Hefei, China
e-mail: wangl@ustc.edu.cn

J. Liu
e-mail: ljsmile@mail.ustc.edu.cn

Y. Zhang
e-mail: yuncan@mail.ustc.edu.cn

Z. Wang
e-mail: zhenfaw@mail.ustc.edu.cn

S. Wang
e-mail: wsong@ustc.edu.cn

© Springer Science+Business Media Singapore 2016
J.J.(Jong Hyuk) Park et al. (eds.), *Advances in Parallel and Distributed Computing and Ubiquitous Services*, Lecture Notes in Electrical Engineering 368,
DOI 10.1007/978-981-10-0068-3_8

1 Introduction

Contents are retrieved directly by their names in CCN, changing the original host-host model in IP. Ubiquitous in-network caching is a key feature of CCN. Clients can hit contents at the intermediate routers instead of source providers. It increases the overhead of network storage, but effectively reduces the response time. Limited storage and the rapidly increasing content quantity have conflicts [1]. Many articles have gradually given their exploration of caching management. Liu et al. [2] propose APDR. The Interest packet collects information of each node along the path, and the responder decides where and what to cache according to these information. The authors in [3] use markov-chain to establish mathematical model of CCN storage and select the optimal placement to reduce the cache redundancy. Rossi et al. [4] indicate that topology and multi-path effect caching performance. Psaras [5] proposes a caching mechanism based on probability. However all schemes above improve cache performance in CCN, they don't consider the importance of topology and router node.

The router in CCN has the built-in content store module (CS) to cache contents. Rossi et al. [6] show that more cache space in the core nodes can improve the performance, and first propose cache allocation deployment scheme. The authors in [7] introduce a two-steps method to get accurate optimal solution of content placement under a total storage budget for the network, providing the reference for network operators on how to deploy cache capacity optimally.

Content-centric networking based on OpenFlow has become an important direction [8, 9]. OpenFlow is a mainstream SDN protocol. Protocol-Oblivious Forwarding (POF) [10] is the extension of OpenFlow. POF controller issues all-purpose instructions to the POFswitch, and forwarding devices support existing and self-defined protocols. So POF can support IP routing and content routing. We have realized a CCN prototype using POF (called SDCN), which doesn't rely on the IP.

The reminder of this paper is organized as follows. In Sect. 2, we describe SDCN architecture, and introduce the concept of Node Value. We also discuss how to realize hierarchical cache allocation. In Sect. 3, we introduce a cache decision policy based on node value. Implementation and analysis of simulation are in Sect. 4. The conclusion is made in Sect. 5.

2 Hierarchical Cache Model

We introduce SDCN architecture and use the Node Value to describe the importance of router node. Then we construct a hierarchical cache model based on the node value.

2.1 Software Defined Content Network Architecture

SDCN architecture consists of two planes, as shown in Fig. 1. Controller Plane issues instructions to the POFswitches according to the demand of users, and controls the forwarding of the switches. It also monitors the status of switches and collects network information for the reasonable resource allocation. Forwarding Plane consists of POFswitches and CCN server nodes. POFswitch does not have built-in cache model, so we need a server node to realize the CS, PIT, FIB. In general, we can treat a POFswitch and a CCN service node as a complete CCN router node.

2.2 Node Value

The importance of node is usually represented as node centricity metric. In the analysis of network, the node centricity metric tends to be used as the minimum path length, because it will set a higher value for the center node which bears higher traffic.

Node centricity: we use D_i as the centricity of *node i*, which is defined as the reciprocal of the sum of shortest path distance to other nodes by Eq. (1), where N is the number of nodes in network. We take normalization process of D_i to process the data efficiently by Eq. (2). The node with higher D_i is near the virtual center of network.

$$D_i = 1/ \sum_{k=1,k\neq i}^{N} d(i,k) \tag{1}$$

$$D_i = (D_i \times N)/ \sum_{i=1}^{N} D_i \tag{2}$$

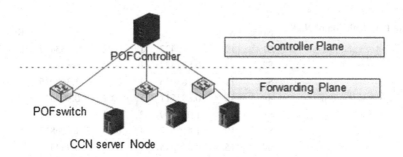

Fig. 1 SDCN architecture

Node centricity may not be enough accurate, because they only consider indirect connection instead of the direct connection. In order to describe the importance of the nodes more accurately, we take the node connectivity into account.

Node connectivity: L_i is the number of nodes connected to *node i*.

Using the above two parameters to describe the importance of node accurately, we introduce a new concept to represent the tradeoff between them—**Node Value**. The Value is defined as $V_i = D_i + \alpha \times L_i$, where α is the tradeoff between Node centricity and Node connectivity. It varies with the network scale. Note that the influence of the Node centricity is much more than the Node connectivity, because Node connectivity improves Node centricity's limitations.

2.3 Hierarchical Cache Model

In large-scale network, nodes with higher value handle more traffic and need more cache space. Because of the expensive hardware, it is important that how to allocate cache space under certain total cache space C. In SDCN, controller can easily access topology and then calculate each Node Value. According to node Value, hierarchical cache model is built, where nodes in higher layer have more space.

We roughly show the calculation of node value and hierarchy division in the sample topology like Fig. 2. For S1, the sum of the shortest path to others is $2 + 1 + 3 + 2 + 2 + 3 + 4 = 17$. The values of other nodes are calculated like this. The results are shown in Table 1, and α is 0.02.

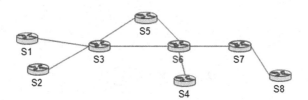

Fig. 2 Simple topology example

Table 1 Calculation of Node Value

Node	Sum	D_i	L_i	$\alpha \times L_i$	V_i	Rank
S1	17	0.833422	1	0.02	0.853422	6
S2	17	0.833422	1	0.02	0.853422	7
S3	11	1.288015	4	0.08	1.368015	2
S4	16	0.885511	1	0.02	0.905511	5
S5	13	1.089859	2	0.04	1.129859	3
S6	10	1.416817	4	0.08	1.496817	1
S7	15	0.944545	2	0.04	0.984545	4
S8	20	0.708408	1	0.02	0.728408	8

Table 2 The allocation of total cache space C according to the layer

	Cache of layer	Nodes	Cache of node
Layer1 (10 %)	C/3	S6	C/3
Layer2 (30 %)	C/3	S3, S5	C/6
Layer3 (60 %)	C/3	S1, S2, S4, S7, S8	C/15

The network with 8 nodes is divided into 3 layers (the number of layers is according to the scale) to show hierarchy division. The total cache space C is divided equally into each layer as C/3, and each node of each layer further divides the C/3 space equally, as shown in Table 2.

3   Cache Decision Strategy Based on Node Value

Based on the hierarchical model above, we propose cache policy NVD for SDCN. Nodes with high value tend to deal with more traffic and have larger cache space. NVD selects the node with highest value along the path to cache the popular content, and lowest one to cache unpopular instead of caching on every node. Content popularity has no exact model and unified definition, so the popularity evaluation should be contained in the future work. We assume the content popularity is known, defined as request probability $P \in (0,1)$. Frequent requests of popular contents may cause more and more copies of popular content which increase redundancy, and less and less unpopular contents which reduces content diversity in the network.

We set a content popularity threshold T in our NVD strategy, the responder compares the value of each node along coming path. The content whose popularity is greater than T will be cached on the node with the highest value, while the content whose popularity is less than T will be cached on the node with the lowest value, in order to ensure the diversity of the network.

4   Experiment and Analysis

SDCN collects network topology through controller and realizes the hierarchical cache model. Due to the limited hardware resource, a topology with 50 nodes is built under Mininet, consisting of two core nodes connected with some random tree-topology with maximum degree of 4. We set this topology into four layers. The total storage budget is 600 objects, and we place 1000 files randomly. Content popularity is modelled with a Zipf distribution and Content requests are modelled as Poisson process. The cache replacement policy is based on popularity [11].

We use $P\_hit = \sum hit / \sum request$ to describe the performance of network. $\sum hit$ is the number of cache hit at one node, and $\sum request$ is the number of requests it receives. The simulation is executed in four scenarios: LCE and NVD with hierarchical cache model (LCE_Model and NVD_Model), while LCE and NVD without model. As shown in Fig. 3, average P_hit of each layer with hierarchical cache model is higher than ones without model. It indicates that hierarchical cache model can increase cache hit ratio. The results in Fig. 4 show P_hit of each node, the X axis is rank of their values. Although P_hit of layer1 (node 1–5) is improved, it still less than layer2 (node 6–15) and layer3 (node 16–30). It may be caused by frequent requests and replacements at nodes with high value. We sample content request, and the proposed NVD gets the average request hops of 4.56, less than 4.76 under LCE.

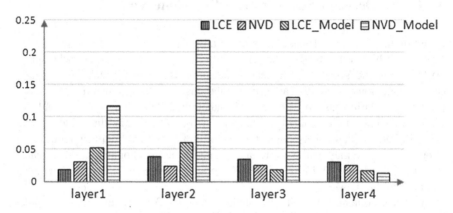

Fig. 3 The average P_hit of each layer in four scenarios

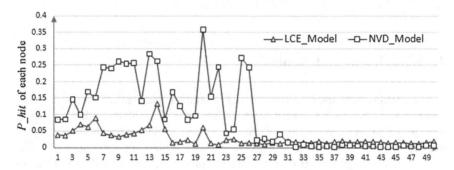

Fig. 4 P_hit of each node using LCE_Model versus NVD_Model

5 Conclusion

Previous researches on CCN over SDN do not focus on the heterogeneity of node caching. We built a SDCN architecture and put forward the concept of node value to describe its importance. The node value is calculated to construct the Hierarchical cache model for cache space allocation. NVD puts the popular content on the nodes with higher value. The simulation shows proposed hierarchical cache model and cache policy can improve the performance of network caching, including hit ratio and average hops of each request. For further research, the study of node value considering more information such as link bandwidth will be required.

Acknowledgements This work is supported by the "Strategic Priority Research Program" of the Chinese Academy of Sciences (XDA06030900).

References

1. Min EL, Chen Z, Hongfeng XU, Liang Y (2012) Research Progress of Content Center Network. J. Netinfo Security. 2:6–10
2. Liu WX, Shunzheng YU, Cai J, Gao Y (2013) Scheme for cooperative caching in ICN. J Softw 24:1947–1962
3. Muscariello L, Carofiglio G, Gallo M (2011) Bandwidth and storage sharing performance in information centric networking. In: Proceedings of the ACM SIGCOMM workshop on information-centric networking, ICN'11. ACM, Toronto, pp 26–31
4. Rossi D, Rossini G (2011) Caching performance of content centric networks under multi-path routing (and more). Telecom ParisTech
5. Psaras I, Chai WK, Pavlou G (2012) Probabilistic in-network caching for information-centric networks. In: Proceedings of the second edition of the ICN workshop on information-centric networking, pp 55–60
6. Rossi D, Rossini G (2012) On sizing CCN content stores by exploiting topological information. INFOCOM Workshops, pp 280–285
7. Wang Y, Li Z, Tyson G, Uhlig S, Gaogang X (2013) Optimal cache allocation for content-centric networking. In: 21st IEEE international conference on network protocols (ICNP). IEEE Press, Goettingen, pp 1–10
8. Blefari-Melazzi N, Detti A, Mazza G, Morabito G, Salsano S, Veltri L (2012) An openflow-based testbed for information centric networking. In: Future network and mobile summit. IEEE Press, Berlin, pp 4–6
9. Veltri L, Morabito G, Salsano S, Blefari-Melazzi N, Detti A (2012) Supporting information-centric functionality in software defined networks. In: IEEE international conference on communications (ICC). IEEE Press, Ottawa, pp 6645–6650
10. Song H (2013) Unleash the power of SDN through a future-proof forwarding plane. In: Proceedings of the second ACM SIGCOMM workshop on hot topics in software defined networking, pp 127–132
11. Bernardini C, Silverston T, Festor O (2013) Cache management strategy for CCN based on content popularity. Lect Notes Comput Sci 7946:92–95

Interoperation of Distributed MCU Emulator/Simulator for Operating Power Simulation of Large-Scale Internet of Event-Driven Control Things

Sanghyun Lee, Bong Gu Kang, Tag Gon Kim, Jeonghun Cho and Daejin Park

Abstract Internet of Event-Driven Control Things (IoEVCT), the large-scale event-based control systems based on the Internet-of-Things (IoT) can be composed of a series of sensors, controllers, and actuators. It is wirelessly connected by the internet of events between sensors and actuators. The perceptible objects are considered as not only a time-based, controlled system at the micro level, but also as an event-driven, controlled system at the macro level of large-scale IoT. This paper introduces our initial effort to implement a newly designed simulation framework for quantitative power measurement when a large number of sensors and actuators are connected wirelessly to control certain plants. Power consumption of the whole system can be divided into power consumed by a microcontroller unit (MCU) in sensors and actuators. Using the proposed architecture, we can range from the low-level viewed power via an MCU to the whole power of the large scale of control things. We try to substitute each physical MCU in the sensor and actuator devices with the emulation based on an instruction set simulator (ISS) to reduce the physical cost caused by experiments in real environment. Using the proposed simulation framework, our study shows a possibility that the sensing, controlling, and actuating process in the wirelessly connected control systems over the large-scale IoT can be analyzed in terms of the power consumption, which is affected by various environmental causes around IoT.

Keywords Event-driven large-scale simulation · Interoperation of hardware and software · Internet-of-things · Distributed control system

S. Lee · B.G. Kang · T.G. Kim
Korea Advanced Institute of Science and Technology (KAIST),
Bukgu, Republic of Korea

J. Cho · D. Park (✉)
School of Electronics Engineering, Kyungpook National University,
80 Daehakro, Bukgu, Daegu 702-701, Republic of Korea
e-mail: boltanut@knu.ac.kr

© Springer Science+Business Media Singapore 2016
J.J.(Jong Hyuk) Park et al. (eds.), *Advances in Parallel and Distributed Computing and Ubiquitous Services*, Lecture Notes in Electrical Engineering 368,
DOI 10.1007/978-981-10-0068-3_9

1 Introduction

The large-scale event-based control systems based on IoT [1] are made up of sensors, controllers, and actuators. Most control systems implemented by computerized programs are based on the fact that time step sizes between controller operations are equal. This means the controller runs periodically with fixed sampling-time. Thus, the mentioned control system is called a time-based control system. The system is executed continually regardless of whether the system output is achieved to a set point.

On the other hand, event-driven control systems activate the controller only when an event, including something important, occurs. Thus, event-driven control systems are more efficient than time-based ones in that the power consumed by MCU for operating the controllers can diminish [2, 3]. To find events in the event-based control system, we make the most of event detector to determine a time when the controller wakes up from its sleep state and begins to work. Conditions for triggering the events are various such as LC, ILC, LP, ILP, and EN (Fig. 1).

In these event-based sampling strategies, LC method is used to keep down the complexity comparing with time-based control system. In LC method, when the difference between the current error and error the last time that condition was true is greater than a certain threshold. Event detector in the event-based control system plays a role in performing the aforementioned. Event detector takes the error at the current time, and generates an event if satisfying below condition, i.e., Eq. 1.

$$|e_{cur} - e_{last}| \geq e_{th} || h_{nom} \geq h_{max} \tag{1}$$

The second inequality in Eq. 1 is used to a simple safety measure. As we examine two control systems in terms of an event sequence, unlike the event sequences of time-based control systems with the constant time step between adjacent events, until the output of the plant reaches the desired value, event

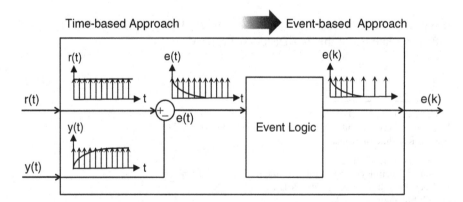

Fig. 1 Event-detector architecture for event-based control

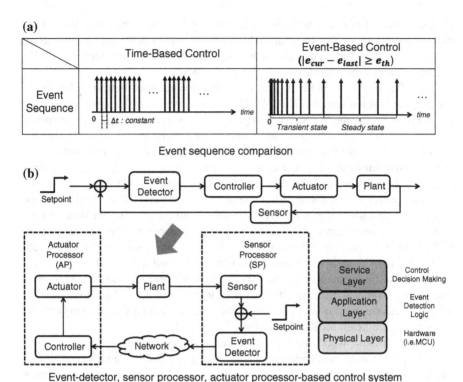

Event sequence comparison

Event-detector, sensor processor, actuator processor-based control system

Fig. 2 IoT devices and control systems represented by the view of internet-of-events

sequence will be generated densely and then sparsely after the output is almost same as the set point in Fig. 2a.

The efforts to decrease the number of activating the controller logic have been studied as a new topic in the control community by the name of event-based control [4, 5]. Such the event-based control can be expressed in the form of an error event passes between smart-sensor and smart-actuator in the IoT environment [6].

2 Research Motivation

Figure 2b represents the block diagrams of the event-based control systems combining sensor and event detectors. The novel device block is called a sensor processor (SP) [7] and actuator processors (AP) can be made by integrating an actuator with a controller. Due to being separated between SP and AP, wired or wireless networks can be formed between SP and AP.

SPs and APs can be located in certain plants, depending on the number and scale of plants. SPs in one of the plants can be wirelessly connected with APs in the other plants, so packet data can be transmitted between SPs and APs even though both plants are different. The original architecture of the system, which consists of sensors, actuators, and certain plants, is called the wireless sensor network (WSN). Figure 3a shows the overall architecture of the WSN.

The difference between the internet of event-driven control things (IoEVCT) and the WSN is that the sensor in the IoEVCT not only performs the signal conversion passively as one in the WSN, but also processes the converted signal to determine whether service should be provided to the users. Therefore, we can see that the IoEVCT is the extended version of the WSN with added processor units in the conventional sensor and actuator in order to offer various services such as control.

When the SP transmits the data into the AP in the IoEVCT environment, the operating power can be dissipated due to many factors such as storing the external input, executing some logics for event detection, and transmitting the data into AP in SP. In the same manner, AP consumes power via processing the received packet. If the source supplied to SP and AP is so highly constrained that we cannot use them for a long time, both SP and AP cannot be applied in the real environment.

In order to find some meaningful context such that both SP and AP can work at low-level power and maintain their own characteristics for a long time, we are going to perform the power simulation in the large-scale IoEVCT environment and extract the significant results to the latter.

As our previous work, the low-power sensor processor architecture [7] shows that based on the event-driven signal processing approach is an efficient design method in terms of power consumption. We focused on the event-driven sensor signal processing in the single MCU device, then our research is extended to cover the actuator process in the event-driven way, especially for the large-scale connection of the sensors and actuators as a main topic of this paper.

This paper as a work-in-progress is organized as follows: Sect. 3 presents the proposed architecture for performing power simulation in the IoEVCT environment. Section 4 suggests what to take into consideration in implementing.

3 Proposed Architecture

The proposed architecture for the IoEVCT simulation using an instruction set simulator (ISS) instead of real hardware platforms for quantitative analysis like power dissipation is described in Fig. 3b as the view of the simulation bus architecture.

Actually, our previous work focused on the power reduction in the single sensor MCU, shown in right-side of Fig. 3b. Using the power model in the single MCU, our research focuses on providing quantitative estimation of the macro-level power consumption for the large-scale connected IoT devices.

The reason why we will use an ISS including the peripherals of MCU rather than real MCU hardware is that we can perform the large-scale simulation without environmental constraints and find defaults during simulation while preventing

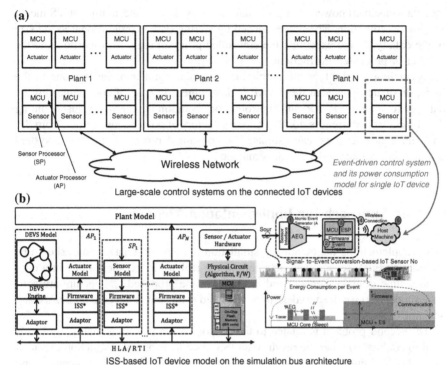

Fig. 3 IoT devices and control systems represented by the view of internet-of-event

hardware damage. Also, we can reduce the development cost required by hardware such as SP and AP.

The proposed simulation architecture is composed of two parts: one is the sensor/actuator part; the other is the control part. Because the ISS is a software model unlike a real MCU, sensor and actuator devices are replaced by software models such as the Discrete Event Specification (DEVS) model. An ISS performs the embedded-program logic like firmware to process the data occurring when the sensed data come in. In the case of SP, event-detection logics are carried out whether or not the event should be generated so as to transmit the information into a controller.

The function of an ISS in AP is to deliver the controller output to the plant. The control part is made up of the DEVS model to present the services. Services are provided in the order of earlier-arrived events. To synchronize the logical time and exchange the data easily, when various simulators, which are comprised of model and simulation engines, participate in the inter-operation, we are going to use high-level architecture (HLA)/real time infrastructure (RTI) as standard.

Factors of dissipated power in overall systems can be enumerated as follows: storage via SP/AP or controller, embedded-program logic in SP/AP, packet transmission through the wireless network, controller operations, and so on. To measure the total consumed power during simulation, we will apply the atomic DEVS model for power estimation. The power caused by storing data is proportional to the size of the data type, and executing the embedded-program logic is proportional to the number of processed instructions.

Depending on the type of instruction and total number of instructions in the logic, dissipated power in the SP/AP is determined. In case of communicating by data packet among SP, AP, and controller, or operation by controller, just counting the number of occurrences do. By gathering factors of power consumption from SP, AP, and controllers, we can estimate the consumed power of the whole system through the atomic model during simulation.

4   Our Approach for Implementation Method

When implementing the mentioned simulation framework, related issues should be considered. First, sensors and actuators in the real world get their value in a continuous time domain. Sensors and actuators in the proposed architecture are software models, so we need to determine the fixed time-step to execute the sensor and actuator.

The second problem is related to the purpose of an ISS. In actuality, an ISS is used to confirm whether the result of an ISS is same as one of instruction executed on an MCU. Thus, many ISSs provide only the results of an MCU, memory, and register. However, when we use real hardware instead of an ISS, other peripherals can be used such as pulse width modulation (PWM). Therefore, we should implement this additionally in case other functions of peripherals are required during simulation. Also, an ISS should be able to be served I/O functions. The reason is that most ISSs are focused on the result of execution logic when the external inputs are already given.

The third problem is relevant to event detection logic. Before transmitting the important context extracted from an event detector, conversion logic from signal to event is needed. The logic is called Signal-to-Event Converter (S2E), shown in Fig. 4, which was introduced in our previous work [7], because for a controller in a service part specified by the DEVS model to adjust the data type from analog signal to event retaining certain value, S2Es are necessary. Also, when we want to modify the event detector, we have to do reprogramming processes to update the logic.

In real hardware, we perform cross-compiling in a standard-alone PC, and then send the compiled-program into the target board to run programs on the target. Although we reset the target board after the program is written into memory, program data stored into memory is not removed until the reprogramming process occurs. So, we should cope with the situation by altering the event-detection logic during simulation.

The fourth problem is always involved with inter-operation. As mentioned before, each SP, AP, and controller can be regarded as a simulator, which consists of model and simulation engines. To exchange the data between different

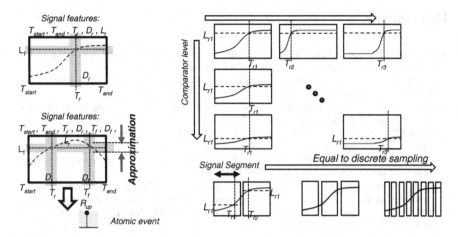

Fig. 4 Signal-to-event converter for event-driven signal representation

simulation adaptors, which means a simulator participating in inter-operation, we need to know what kinds of data will be exchanged. To make the data conversion simple via the simulation bus, we should add the adaptor between the single simulator and HLA/RTI simulation bus standard.

The last problem can occur when transmitting the packet from controller to AP. As a controller does not include the data about where the controller output is sent, we should keep in mind that before sending the detected event to controller, events should contain such as information about destination.

5 Conclusion

With the sensor/actuator emulation using an ISS instead of a real MCU in the proposed architecture, our research objective is to obtain event-driven contexts for the quantitative factors such as power dissipation, throughput, latency, etc., which are represented with the internet of events. The proposed simulation framework can reduce risk from using real devices, hardware faults, development costs, and the constraints of the environment. Using the proposed framework, the sensor, controller, and actuator processes in wirelessly connected control systems can be analyzed in terms of a specific performance index like power consumption.

Acknowledgements This research was supported by the Basic Science Research Program through the National Research Foundation of Korea (NRF) funded by the Ministry of Education (2014R1A6A3A04059410).

References

1. Lazarescu M (2013) Design of a WSN platform for long-term environmental monitoring for IoT applications. IEEE J Emerg Select Topics Circuits Syst 3(1):45–54
2. Tsividis Y (2010) Event-driven data acquisition and continuous-time digital signal processing. In: Custom integrated circuits conference (CICC). IEEE, pp 1–8
3. Tsividis Y (2010) Event-driven data acquisition and digital signal processing: a tutorial. IEEE Trans Circuits Syst II Exp Briefs 57(8):577–581
4. Årzén KE (1999) A simple event-based PID controller. In: Proceedings of 14th IFAC world congress, vol 18, pp 423–428
5. Durand S, Marchand N (2009) Further results on event-based PID controller. In: Control conference (ECC), 2009 European, pp 1979–1984
6. Grüne L, Jerg S, Junge O, Lehmann D, Lunze J, Müller F, Post M (2010) Two complementary approaches to event-based controlzwei komplementäre zugänge zur ereignisbasierten regelung. at-Automatisierungstechnik Methoden und Anwendun- gen der Steuerungs-, Regelungs-und Informationstechnik 58(4):173–182
7. Park D, Cho J (2014) Accuracy-energy configurable sensor processor and IoT device for long-term activity monitoring in rare-event sensing applications. Sci World J

The Greedy Approach to Group Students for Cooperative Learning

Byoung Wook Kim, Sung Kyu Chun, Won Gyu Lee and Jin Gon Shon

Abstract For effective cooperative learning organize group is important. Member in the group to the interaction between the group members are to be composed of heterogeneous. However, the average ability of the group needed to solve a given task to a fair evaluation in cooperative learning should be similar to each other between the groups. In this paper, we propose greedy approach to find partitions with high homogeneity in a group and high heterogeneity between groups.

Keywords Grouping student · Partitioning · Team formation

1 Introduction

Cooperative learning is a learning style. In this style regular group members set the learning goals of the joint and resolve the problem with the obligations in the same position in order to reach that goal that member out to yield beneficial results for all of the learning type to highlight the evaluation of the joint on the result [1].

B.W. Kim · S.K. Chun
Department of Computer Science Education, Korea University,
Seoul, Republic of Korea
e-mail: byoungwook.kim@inc.korea.ac.kr

S.K. Chun
e-mail: sungkyu.chun@inc.korea.ac.kr

W.G. Lee
Department of Computer Science and Engineering, Korea University,
Seoul, Republic of Korea
e-mail: lee@inc.korea.ac.kr

J.G. Shon (✉)
Department of Computer Science, Korea National Open University,
Seoul, Republic of Korea
e-mail: jgshon@knou.ac.kr

© Springer Science+Business Media Singapore 2016
J.J.(Jong Hyuk) Park et al. (eds.), *Advances in Parallel and Distributed Computing and Ubiquitous Services*, Lecture Notes in Electrical Engineering 368,
DOI 10.1007/978-981-10-0068-3_10

83

Recently, interests in cooperative learning become very higher. The reason why interest is cooperative learning has several advantages that promote academic achievement, student interest caused, social skills development, providing a variety of teaching strategies.

The first starting point to practice the cooperative learning is to form organizing a group. Traditional group learning uses a homogenous group. If a group together children of the same character classes have more academic achievement deviation with progress class activity. In contrast, the basic principles that make up the group cooperative learning is a heterogeneous group. In other words, a group member is to consist of a group such as different gender or personality, learning style among the other children. The reason for this is that the heterogeneity is active the interaction group peer [2].

Recently, some studies are underway to divide the students into their group in the computational aspects [3, 4]. Majumder et al. [3] take efficient group organization as group member's skills needed by the task and communication ability to communicate each other members. Agrawal et al. [4] divide into group members, leaders and followers on the basis of average ability group members. He assume that follower are being helped to learn from other members, leaders improve learning while help other members. Study focuses on the benefits that they gain followers. We proposed an algorithm to divide the group so that maximize followers' benefit, when divided the students into l groups.

2 Related Work

2.1 Cooperative Learning

Cooperative learning is a teaching method which students has different skills work together to achieve same goal in small groups for learning activities in the learning group when reward based on grades of the group [5]. The essence of cooperative learning groups is defined as a series of teaching- learning process that assigned a goal to reach a common goal achievement, increase the average grade of the group, and compensate the entire members by examining the quantity and quality of grades according to a predetermined criterion [2].

There are some methods for organize a group, like teacher centered method, student centered method, random process method, homogeneous group method. Teacher centered method is to organize evenly group according to several criteria, such as grades or personality driven by teacher. Student centered method is teacher elect a student to leader of a group, then leader is adjusted to pull the rest of the students in his (or hers) group. Random process method is that comprise the students into groups randomly by using a piece of picture cards or etc. Homogeneous group method is to organize groups among the students who have the same theme or concerns like project or lesson. Because each method has the strengths and weaknesses, method is used to choose according to the characteristics of the students' situation or section curriculum.

2.2 Finding a Team of Experts

Team Formation has been studied in the operation research community [3]. Recently, some studies of the finding a team of experts to complete a given project have been made active. These studies propose a method to measure whether the communication cost, only to find the members that meets the skills required for the task in groups composed of individuals with diverse skill to work together effectively as groups. Communication cost was measured using a social network.

The above studies' purpose is to find the single best team. So, it is different from the dividing entire group into a number of groups. Since only it covers categorical variables for determining whether group members have a skill for solving the task is not suitable for dealing with numerical data variables.

2.3 Grouping Students with Ability

In the computational perspective, studies are to divide students into groups according to their ability [4]. The ability of learners to a particular subject was assumed to $\theta \in [0, 1]$. An average ability of group with members is regarded as collective ability of a group, at a member who has lower ability than a collective ability of a team is defined a follower and the others members whose ability is more than a collective ability of a team, collective members is defined leaders. In a group, students can improve the ability to interact with and cooperate with other members. These are assumed that Follower will have some gains may receive help learning from other members, leaders will have some gains to improve the learning, giving help to other members. The study focused on the follower's gain that they get. When divided the students into l groups, that proposed an algorithm to divide the group that the maximize benefit is gained by followers. In the above study, the ability of the learner deals only with one particular field. In the group's purpose to maximize benefit is gained by followers.

3 Framework

3.1 Preliminaries

We assume that n students $S = \{s_1, s_2, ..., s_n\}$. Each student s_i has ability $\theta_i \in [0, 1]$ in a specific domain. We assume $K = \{k_1, ..., k_m\}$ to be m skills. The student can be represented by a set of tuples consisting of the skill and ability of the skill.

For example, if s_i has abilities θ_{i1}, θ_{i2} with regard to k_1 and k_2, the student can be represented by $s_i = \{(k_1, \theta_{i1}), (k_2, \theta_{i2})\}$, $k_1, k_2 \in K \backslash \{k_1, k_2\}$ is a abilities that s_i does not have. F_s is a function that maps a set of students to a set of skills. $F_s(s_i) = \{k_1, k_2\}$. F_θ returns a set of abilities that students has. $F_\theta(s_i) = \{\theta_{ik}, \theta_{it}\}$. We assume $A = \{a_1, \ldots, a_o\}$ to be o attributes. The student is represented by the attribute vectors.

A task T is a set of ability needed to perform the cooperative learning, $T \subseteq K$. That one student has the skills required by a given task, we measure the student cover.

$$C_S(s_i, T) = \frac{|F_s(s_i) \cap T|}{|T|} \tag{1}$$

N students are divided into groups $G = \{G_1, \ldots, G_l\}$. A_G presents a set of skills that students have been included in the group.

$$A_{G_j} = \bigcup_{s \in G_j} F_s(s) \tag{2}$$

That one group has the skills required by a given task T, we measure the group cover as follows.

$$C_T(G_i, T) = \frac{|A_{G_j} \cap T|}{|T|} \tag{3}$$

Definition 1 Given a task T, if $C_T(G_i, T) = 1$ then G_i covers task T.

Team members in the cooperative learning should be heterogeneous with each other (Fig. 1). For example let that a group is composed of four students. There are a property representing the learning style is being and different each other four student can be called heterogeneous. Property to determine the student's heterogeneity is assumed that predefined as a categorical variable. Heterogeneity of the property is determined by entropy of information theory-based.

Fig. 1 The concept of heterogeneity of a group and homogeneity between groups

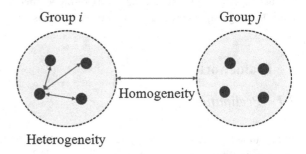

The entropy of the properties of a group is determined using Eq. (4).

$$H(a_i) = -p \cdot \log_2 p \tag{4}$$

where a_i is the i attribute and p is the probability of each attribute in a team.

Heterogeneity of the one group is the sum of the entropy for every attribute as follows.

$$H_e(G_k) = \sum_{i=1}^{o} H(a_i) \tag{5}$$

where G_k is the k group and o is the number of attributes in a data set.

Heterogeneity of all groups is the sum of the heterogeneity of each group as follows.

$$H_e(G) = \sum_{k=1}^{l} H_e(G_k) \tag{6}$$

where l is the number of groups.

When organize a group for the cooperative learning, each group must have similar abilities. We need to define the average ability $\hat{\theta}_G$ of a group. The average ability of a group is the average ability of members in the group [6].

$$\hat{\theta}_G = \frac{1}{|G|} \sum_{i=1}^{|G|} F_\theta(s_i) \tag{7}$$

where $|G|$ is the number of member in a group.

The difference of ability between the two groups can be determined as follows

$$D(G_i, G_j) = \left| \hat{\theta}_{g_i} - \hat{\theta}_{g_j} \right| \tag{8}$$

Measuring the total homogeneity of the group is the summation of difference all pairwise groups.

$$H_o(G) = \left| \sum_{G_i, G_j \in G} D(G_i, G_j) \right|, i \neq j \tag{9}$$

3.2 Problem

In this paper, we are interested in finding partitions with high homogeneity in a group and high heterogeneity between groups. The formal definitions of this problem is given below.

Problem 1 (*Team Formation for Cooperation Learning*) Given integer k, a set of n students $S = \{s_1, s_2, \ldots, s_n\}$, and task T, find a partition of S into groups $G_1, \ldots, G_l$, where the size of each group is k and $H_e(G)$ and $H_o(G)$ are maximized.

4 Algorithm

As shown as [4], the partition problem that n students partition l groups is NP-complete. To solve this problem, we adopt the greedy approach. The pseudocode of the greedy approach is shown in algorithm.

Greedy approach algorithm

Input: Set of students $S = \{s_1, s_2, \cdots, s_n\}$ with sorted total abilities $\theta_1 > \theta_2 > \cdots > \theta_n$, group size k, and number of groups r ($2 \leq r \leq n/2$).
Output: Group G.
1: $G = \emptyset$
2: Randomly choose r students as s_i ($i = 1, \cdots, r$)
3: $S \leftarrow S \setminus \{s_i\}$
4: **for** i=1 to r **do** /*assign r student to each group */
5:   $G_i \leftarrow s_i$
6: **for** j=1 to (n-r) **do**
7:   **for** i=1 to r **do**
8:     $G_i \leftarrow s_j$ /*assign a student to a group temporary */
9:       $H_i \leftarrow H_e(G) + H_o(G)$ /*calculate homogeneity and heterogeneity*/
10:    $G_i \leftarrow G_i \setminus \{s_j\}$ /*remove a student from G_i*/
11:   Assign $G_i \leftarrow s_j$ with the highest H_i among G
12: **return** G

---

5 Conclusions

In this paper, we proposed the greedy approach for finding partitions with high homogeneity in a group and high heterogeneity between groups. We define the problem of the Team Formation for Cooperation Learning formally. Using this problem definitions, we measured homogeneity in a group and heterogeneity between groups. In the future, we would like to conduct experiments to measure our algorithm proposed in this paper.

References

1. Terenzini PT, Cabrera AF, Colbeck CL, Parente JM, Bjorklund SA (2001) Collaborative learning vs. lecture/discussion: students' reported learning gains. J Eng Educ 90:123–130
2. Johnson DW, Johnson RT (1989) Cooperation and competition: theory and research. Interaction Book Company, Edina
3. Majumder A, Datta S, Naidu KVM (2012) Capacitated team formation problem on social network. In: KDD, pp 1005–1013
4. Agrawal R, Golshan B, Terzi E (2014) Grouping students in educational settings. In: Proceedings of the 20th ACM SIGKDD international conference on Knowledge discovery and data mining (KDD'14). ACM, New York, pp 1017–1026
5. Slavin RE (1980) Effects of individual learning expectations on student achievement. J Educ Psychol 72:520–524
6. Mislevy RJ (1983) Item response models for grouped data. J Educ Stat 8(4):271–288

Secure Concept of SCADA Communication for Offshore Wind Energy

Seunghwan Ju, Jaekyoung Lee, Joonyoung Park and Junshin Lee

Abstract In the power industry, the focus has been almost exclusively on implementing equipment that can keep the power system reliable. Until recently, communications and information flows have been considered of peripheral importance. However, increasingly the Information Infrastructure that supports the monitoring and control of the power system has come to be critical to the reliability of the power system. Communication protocols are one of the most critical parts of power system operations, both responsible for retrieving information from field equipment for sending control commands. We studied the wind energy SCADA system for IEC 61850 communication protocol. We focus on security for reliable SCADA system now. We want to introduce security concept required in SCADA systems.

Keywords Secure SCADA · Communication security · Control system security

1 Introduction

In the power industry, we has been focused that can keep the power system reliable on implementing equipment. Until recently, it have been considered importance to communications and information flows. However, it has come to be critical that supports the monitoring and control of the power system to the reliability.

S. Ju (✉) · J. Lee · J. Park · J. Lee
Korea Electric Power Corporation Research Institute, 105, Munji-Ro,
Yuseong-Gu, Daejeon, Republic of Korea
e-mail: jooseunghwan@gmail.com

J. Lee
e-mail: jklee78@kepco.co.kr

J. Park
e-mail: asura@kepco.co.kr

J. Lee
e-mail: 6lax8e@kepco.co.kr

© Springer Science+Business Media Singapore 2016
J.J.(Jong Hyuk) Park et al. (eds.), *Advances in Parallel and Distributed Computing and Ubiquitous Services*, Lecture Notes in Electrical Engineering 368,
DOI 10.1007/978-981-10-0068-3_11

As the power industry relies increasingly on information to operate the power system, we must now be managed the Information Infrastructure.

The management of the power system infrastructure has become reliant on the information infrastructure as automation continues to replace manual operations, as market forces demand more accurate and timely information, and as the power system equipment ages. The reliability of the power system [1] is increasingly affected by any problems that the information infrastructure might suffer, and therefore the information infrastructure must be managed to the level of reliability needed to provide the required reliability of the power system infrastructure.

In the wind power case, the reliability of the wind power system is increasingly affected by any problems that the information infrastructure might suffer, and therefore the information infrastructure must be managed to the level of reliability needed to provide the required reliability of the wind power infrastructure.

So, we wanted to study that secure concept of SCADA communication for Korea offshore wind farm.

2 Need to Secure Communication

Communication protocols are one of the most critical parts of power system operations, both responsible for retrieving information from field equipment and, vice versa, for sending control commands. Despite their key function, to-date these communication protocols have rarely incorporated any security measures, including security against inadvertent errors, power system equipment malfunctions, communications equipment failures, or deliberate sabotage. Since these protocols were very specialized, Obscurity has been the primary approach of security. However, security by obscurity is no longer a valid concept. Electric power is a critical infrastructure in all nations and therefore an attractive target for attacks. The increasing security threats from rogue individuals and nation states have become particularly evident in the recent *Stuxnet* worm [2–4] and Flame malware attacks [5, 6].

Wind energy systems are cyber-physical systems which combine power system operational equipment with cyber-based control of that equipment. Cyber-physical systems are designed not only to provide the functions that the equipment was developed for, but also to protect that equipment against equipment failures and often against certain types of mistakes.

Security for cyber-physical systems are mostly the same as for purely cyber systems, but there are some important differences.

- **Physical impacts**
  First, attacks can cause physical results, such as power outages and damaged equipment. So the threats are against the functions of these systems, not directly on the data itself. Successful attacks on data not only may affect that data, but more importantly can cause some physical world impact.

- **Cyber-physical protections**
  Secondly, since cyber-physical systems already are designed with many protections against "equipment and software failures", some attacks may already be protected against or may simply invoke existing cyber-physical reactions to mitigate the impact of the attack. These intrinsic mitigations should be utilized and possibly enhanced to meet additional types of threats.
- **Cyber-physical mitigations**
  Thirdly, overall cyber-physical systems are designed to "cope" with "attacks" through fault-tolerant designs, redundancy of equipment, and applications that model the physical systems using the laws of physic. Again, these types of system designs should also be utilized and enhanced to make these systems less vulnerable to malicious attacks.
- **Impacts from security**
  Fourthly, some types of cyber mitigation procedures and technologies can negatively impact cyber-physical systems. Therefore the types of security mitigations must be carefully woven into cyber-physical mitigations to ensure that the primary functionality is maintained, even during attacks.

3 Security Concepts

3.1 Threats

Security entails a much larger scope than just the authentication of users and the encryption of communication protocols. End-to-end security involves security policies, access control mechanisms, key management, audit logs, and other critical infrastructure protection issues. It also entails securing the information infrastructure itself [7, 8].

The threats can be realized by many different types of attacks. The same type of attack can often be involved in different security threats. This web of potential attacks means that there is not just one method of meeting a particular security requirement: each of the types of attacks that present a specific threat needs to be countered (Fig. 1).

3.2 Security Processes

- **Interactions with more stakeholders**
  Utilities must exchange information with many other stakeholders, including other utilities, retail energy service providers, smart-meters at customer sites, widely distributed small generation and storage systems, and many other businesses.

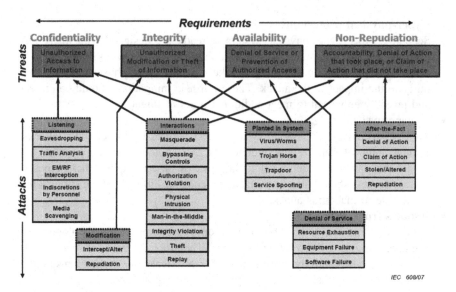

Fig. 1 Security requirements, threats, and attacks, addressed by WG15 [9]

- **Network configurations**
  Although sensitive operational systems are never supposed to be "directly connected with the Internet" or other unauthorized networks, sometimes they are indirectly connected through mis-configurations, handheld devices, and even thumb-drives.
- **Internet-based technologies**
  Utilities increasingly use "open systems", Internet-based technologies, and general consumer products rather than their legacy, one-of-a-kind products. These modern technologies are less expensive and generally more interoperable, but are also more familiar to malicious threat agents who are able to access them and find the inevitable vulnerabilities.
- **Integration of legacy systems**
  At the same time, the existing or "legacy" systems have to be integrated with these more modern systems, often through "gateways" and "wrapping" which lead to their own security vulnerabilities.
- **Increased attraction of the power industry to cyber attackers**
  The power industry, as a Critical Infrastructure that is vital to national security, is subject to the growing sophistication of cyber attackers and to the increasing desire of these cyber attackers to cause financial and/or physical harm the power industry.

4 Apply Security to Wind Energy

Wind energy pose many security challenges that are different from most other industries. In the security industry there is typically a lack of understanding of the security requirements and the potential impact of security measures on the communication requirements of wind energy SCADA.

In particular, the security services and technologies have been developed primarily for industries that do not have many of the strict performance and reliability requirements that are needed by power system operations. The first, preventing an authorized dispatcher from accessing power system substation controls could have more serious consequences than preventing an authorized customer from accessing his banking account. Therefore, denial-of-service is far more important than in many typical Internet transactions. And then, any communication channels used in the power industry are narrowband, thus not permitting some of the overhead needed for certain security measures, such as encryption and key exchanges.

Most systems and equipment are located in wide-spread, unmanned, remote sites with no access to the Internet. This makes key management and some other security measures difficult to implement. Many systems are connected by multi-drop communication channels, so normal network security measures cannot work. Finally, although wireless communications are becoming widely used for many applications, utilities will need to be very careful where they implement these wireless technologies, partly because of the noisy electrical environment of substations, and partly because of the very rapid and extremely reliable response required by some applications (Table 1).

Security measures important to power system operations especially wind energy. Because of the large variety of communication methods and performance characteristics, as well as because no single security measure can counter all types of threats, it is expected than multiple layers of security measures will be implemented (Fig. 2).

Table 1 Specific threats of energy control system

Threats	Description
Carelessness	Passwords on their computer monitors or doors unlocked
Bypass controls	Turn off security measures, Not change default passwords, Everyone uses the same password to access all equipment
Authorization violation	Undertakes actions for which they are not authorized, Masquerade, theft, or other illegal means
Man-in-the-middle	The data is supposed to flow through that this middle equipment is read or modified when it is sent on its way
Resource exhaustion	Equipment is overloaded and cannot perform its functions. Denial of service can become out of control the system

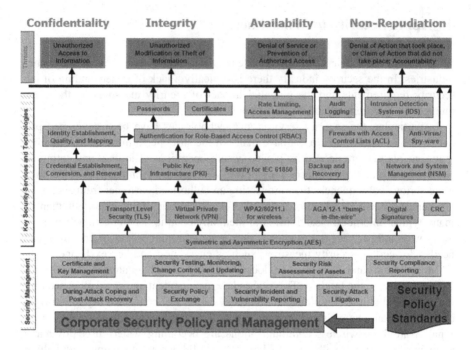

Fig. 2 Security requirements, threats, countermeasures, and management

5   Conclusion

Security entails a much larger scope than just the authentication of users and the encryption of communication protocols.

We are going to study about security technologies such as security for XML files, key management. XML files are used in configuration file representing the wind turbine architectures. An additional security issue requires the development of a use case to analyze how a source of data can identify what data may or may not be made available to other entities in addition to the initial receiving entity.

Acknowledgements This work was supported by the New and Renewable Energy of the Korea Institute of Energy Technology Evaluation and Planning (KETEP) grant funded by the Korea government Ministry of Knowledge Economy (Project No. 20113040020010). We would like to thank KETEP and Korea government MKE for the funding.

References

1. Chen Y, Liu D, Xu B (2013) Travelling wave fault location equipment modeling based on IEC 61850. Dianli Xitong Zidonghua/Autom Electric Power Syst 37(2):86–90
2. IEC 61131-1 ed2.0 (2003) Programmable controllers. Part 1: General information

3. Langner R (2011) Ralph Langner: cracking Stuxnet, a 21st-century cyber weapon. TED. In: TED conferences, LLC
4. Dang B, Ferrie P (2011) 27C3: adventures in analyzing Stuxnet. CCC-TV. Chaos Computer Club e.V
5. Miao X, Chen X (2012) Cyber security infrastructure of smart grid communication system. In: China international conference on electricity distribution, CICED, art no 6508410
6. The short path from cyber missiles to dirty digital bombs. Blog. Langner Communications GmbH. 26 Dec 2010
7. IEC 61850-10 ed2.0 (2012) Communication networks and systems for power utility automation. Part 10: Conformance testing
8. Ke Y-K, Huang P-H, Hsu C-Y, Huang B-Y (2013) Study on communication protocol application in smart grid. Appl Mech Mater 284–87:3235–3239
9. IEC 61400-25-1 ed1.0 (2006) Wind turbines—Part 25-1. Communications for monitoring and control of wind power plants—overall description of principles and models

ASR Error Management Using RNN Based Syllable Prediction for Spoken Dialog Applications

Byeongchang Kim, Junhwi Choi and Gary Geunbae Lee

Abstract We proposed automatic speech recognition (ASR) error management method using recurrent neural network (RNN) based syllable prediction for spoken dialog applications. ASR errors are detected and corrected by syllable prediction. For accurate prediction of a next syllable, we used a current syllable, previous syllable context, and phonetic information of next syllable which is given by ASR error. The proposed method can correct ASR errors only with a text corpus which is used for training of the target application, and it means that the method is independent to the ASR engine. The method is general and can be applied to any speech based application such as spoken dialog systems.

Keywords Automatic speech recognition · Error correction · Neural network

1 Introduction

An automatic speech recognition (ASR) system translates speech input into the correct orthographic form of text. However, ASR system providers occasionally do not provide all necessary components for application development such as the ASR model trainer and the core part of the ASR decoder. Therefore, problems caused by the ASR system must be controlled by tuning its output, so postprocessing may be required to correct the ASR errors.

B. Kim (✉)
Catholic University of Daegu, Gyeongsan, Gyungbuk, South Korea
e-mail: bckim@cu.ac.kr

J. Choi · G.G. Lee
Pohang University of Science and Technology, Pohang, Gyungbuk, South Korea
e-mail: chasunee@postech.ac.kr

G.G. Lee
e-mail: gblee@postech.ac.kr

© Springer Science+Business Media Singapore 2016   99
J.J.(Jong Hyuk) Park et al. (eds.), *Advances in Parallel and Distributed Computing and Ubiquitous Services*, Lecture Notes in Electrical Engineering 368,
DOI 10.1007/978-981-10-0068-3_12

Many previous post-processing methods need parallel corpora which include ASR result texts and their correct transcriptions [1, 2]. Brandow and Strzalkowski [1] suggested a rule-based method; in a training stage, the method generates a set of correction rules from the ASR results and validates the rules against a generic corpus. In a post-editing stage, the set of rules is used to detect and correct the ASR errors. Jeong et al. [2] used a noisy channel model to detect error patterns in the ASR results. The noisy channel model was trained by the parallel corpus.

However, parallel corpora are generally hard to obtain, and they also make error correction dependent on the acoustic environment and on a specific ASR. For this reason, we propose an ASR error management method that needs only correct transcriptions, that is, normal text corpora [3].

Recurrent neural networks (RNNs) are widely used in the ASR field, because powerful to analyzing the sequential data. Robinson [4] applied RNNs to the continuous speech recognition. Graves et al. [5] suggested deep RNNs for speech recognition. For a language model, Mikolov [6] devised an RNN language model (RNNLM) which is powerful to prediction of next word and also powerful to training the word representations. We apply the RNNLM approach to ASR error management, and the RNNs model needs only the corpus that is used for training the application (e.g., dialog system) and that includes only correct sentences. That means that our method is independent of the ASR engine. In the following section, we describe our proposed method. In Sect. 3, we show the experimental results and discuss, and finally in Sect. 4, we conclude.

2 Method

2.1 Overall Architecture

Our proposed method (Fig. 1) consists of two parts: ASR error detection and correction. First, the error detection part detects errors in the input sentence. Next, the correction part replaces or removes words that were identified as errors by the detection part. All the models that are needed to process the method are constructed from only the text corpus that is to train the dialog system.

2.2 ASR Error Detection

ASR error detection is the problem of labeling a word as an error. However, this detection cannot be treated as a supervised classification problem because no parallel corpus that includes ASR results and their transcripts is provided. The errors are essentially detected by voting from each of the detection component modules that independently identify error candidates.

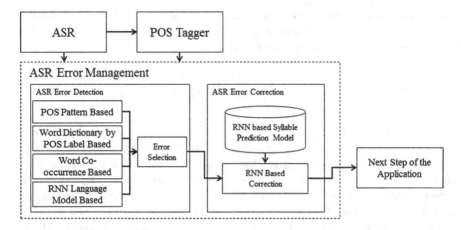

Fig. 1 Architecture of the proposed method

POS Pattern Based Detection An erroneous sentence may have an incorrect POS pattern, such as a grammatical error pattern. With a correct POS pattern, we could detect the erroneous words. POS pattern based error detection model includes several sentence-level POS label sequences. After tagging the ASR output sentence, the system searches for the most similar POS pattern from the model. To find the most similar POS pattern, we use the Levenshtein distance to calculate a similarity score:

$$s = \frac{\text{Levenshtein Distance}(t, p)}{\# \text{ of words of } o}, \tag{1}$$

where t is a POS pattern in an ASR output, p is a POS pattern of the POS pattern model, and o is the ASR output. The lowest scored pattern among all POS patterns in the POS pattern model is selected for error part detection in the ASR output. Aligning the POS label sequence of the ASR output with the selected POS label pattern, any word that does not have a matching POS label in the POS pattern is regarded as an error candidate.

Word Dictionary by POS Label Based Detection Out of vocabulary (OOV) words in the dialog corpus have the possibility of being incorrect words. To construct a word dictionary by POS label, we consider valuable POS labels for the application: i.e., nouns and verbs. If a word in the input sentence is tagged with a valuable POS, the component searches for the word in the dictionary of the tagged POS label. A word that is not present in the dictionary is regarded as an error candidate.

Word Co-occurrence Based Detection Word co-occurrence based detection model includes the target word and its sentence level co-occurring words, which are sorted by co-occurrence frequency.

For each word in the ASR output, a set of co-occurrence words that includes the word itself is constructed by searching the co-occurrence model. The co-occurrence score c_i is calculated by comparing the sets:

$$c_i = \sum_{j \in N} \frac{n(S_i \cap S_j)}{n(S_i)} \times \frac{1}{n(I)}, \tag{2}$$

where S_i is a set of co-occurrence words for word i, N is a set of ASR output words except word i, I is a set of ASR output words, and the function $n(A)$ is the number of elements of A. The numbers of elements of S are equivalent for all i and are determined by a configuration option of the detection component. The words with comparatively low scores in relation to the other words in the ASR output may be possible errors. Then, k words with low c_i are regarded as error candidates. The number of error candidates, k, is determined by a configuration option of the detection component based on the ASR accuracy.

RNNLM Based Detection The RNNLM is trained to generate the word probability distribution given previous context, so the model can be used for evaluation of the appropriateness of each word in an input sentence. The equation of RNNLM score r of the word at position i in the input sentence is

$$\text{RNNLM score } r_i = p(w_i | w_{i-1}, \ldots, w_1), \tag{3}$$

where the probability p is the output of the RNNLM. In the same way as the word co-occurrence based detection, k low scored words are regarded as error candidates.

2.3 Syllable Prediction RNN-Based Error Correction (SPREC)

Before the correction process, the words near the detected erroneous words are also labeled as errors because the neighbor words of the detected erroneous words also have high potential to be incorrect. The error correction method uses a syllable prediction based on RNN. Our method continuously predicts syllables at the detected error position, and the length of the prediction depends on the length of the detected error position. To select a correct word, each generated word replaces detected erroneous word and each revised sentence is evaluated by a word-level likelihood score produced by a language model based on RNN [6]. The sentence with the highest score is selected as the correction.

The syllable prediction network in our method (Fig. 2) has input layer x, syllable context layer h, predicted pronunciation layer p, and output syllable layer y. In syllable position t, the input layer to the network is $x(t)$, the syllable context layer is $h(t)$, the predicted pronunciation layer $p(t)$, and the output syllable layer is $y(t)$. Input layer $x(t)$ is formed by concatenating layer $s(t)$ that represents a current

Fig. 2 RNN for syllable
prediction

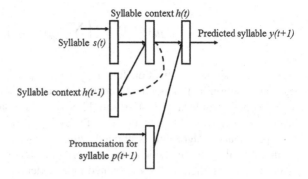

syllable with 1-of-N coding and the previous syllable context layer $h(t-1)$. To predict a syllable in position $t+1$, the layers are calculated as

$$x(t) = s(t) + h(t-1) \tag{4}$$

$$h_j(t) = f\left(\sum_i x_i(t)u_{ij}\right) \tag{5}$$

$$y_k(t+1) = g\left(\sum_j h_j(t)v_{kj} + \sum_l p_l(t+1)w_{kl}\right), \tag{6}$$

where f is a sigmoid activation function and g is a softmax function. The predicted pronunciation layer p is an additional layer that is included for accurate prediction and is provided in two different ways. First, if the position of the prediction $t+1$, provides the pronunciation information, the pronunciation layer represents a confused phoneme sequence of a syllable of the error position $t+1$ and the layer is calculated from the pronunciation confusion matrix [7]. Second, if the position of the prediction $t+1$ cannot provide the pronunciation information, then the pronunciation layer is calculated by the pronunciation RNN. The network for the pronunciation RNN has input layer x_p, pronunciation context layer h_{pn}, and predicted output pronunciation layer p_o. Input layer $x_p(t)$ is formed by concatenating layer $p_c(t)$ which represents a current syllable pronunciation with 1-of-N coding and previous pronunciation context layer $h_p(t-1)$. To predict syllable pronunciation in position $t+1$, the layers are calculated as

$$x_p(t) = p_c(t) + h_p(t-1) \tag{7}$$

$$h_{p_n}(t) = f\left(\sum_m x_{p_m}(t)u_{p_{mn}}\right) \tag{8}$$

$$p_o(t+1) = f\left(\sum_n h_{p_n}(t)v_{p_{no}}\right), \tag{9}$$

where f is a sigmoid activation function. The output layer p_o is activated by the sigmoid function, not the softmax function, because this layer is also an input layer to the output syllable layer y, so p_o should be scaled the same as the syllable context layer h. To train weights u, v and w of the syllable prediction network, a standard back-propagation algorithm is applied with the 1-of-N coding syllable vector to ensure that the output syllable layer represents the next syllable. The syllable pronunciation prediction RNN is trained independently. To train weights u_p and v_p of the pronunciation RNN, a standard backpropagation algorithm is also applied with the 1-of-N coding pronunciation vector to ensure that the output pronunciation layer represents the next syllable pronunciation. To train the RNNs of correction model, weights are initialized to small values as $-0.1 \sim 0.1$. The networks are trained in several epochs. The weights are trained with 0.1 of initial learning rate, and after each epoch the networks is tested on validation data which is the training data. If improvement on the validation data is not significant, then the learning rate is halved and start new epoch [6]. Training process is finished when no significant improvement on the validation data is again [6].

3 Experiments

We evaluated the performance of the proposed error detection and correction method on Korean text. For testing, we prepared a parallel corpus (~ 6500 sentences). The ASR results were generated by an ASR system whose language model was constructed from an open domain corpus with $\sim 300,000$ words and word error rate (WER) of 16.43 %. To train the proposed error detection and correction model, we used a corpus with $\sim 29,000$ sentences that do not include the correct sentences of the test corpus.

3.1 ASR Error Detection

We use a voting method in which a word regarded to be an error must receive the vote from the same number or more than the threshold of the detection components (Table 1).

Table 1 Error detection performance of Korean ASR

Criteria	Precision	Recall	F1-score
POS label pattern	0.501	0.451	0.475
Word dictionary by POS label	0.979	0.110	0.198
Word co-occurrence	0.351	0.510	0.416
RNN language model	0.392	0.443	0.415
Voting (threshold = 1)	0.374	0.773	0.504
Voting (threshold = 2)	0.568	0.442	0.497
Voting (threshold = 3)	0.830	0.372	0.514

The error candidates that get votes from the same number of detection components as the threshold or over are regarded as detected errors

Table 2 Word error reduction rate of SPREC by detection threshold

Criteria	Word error reduction rate (%)
Voting (threshold = 1)	0.27
Voting (threshold = 2)	2.09
Voting (threshold = 3)	8.21

3.2 ASR Error Correction

Our error correction method reduced the word error (Table 2); the best word error reduction was given when the voting voting threshold of the error detection was 3. These results indicate that the increase of the precision of the ASR error detection leads to the exponential level of increase of the word error reduction, because the false detection of the error word causes the false correction even when the error detection has the high recall.

4 Conclusion

In this paper, we proposed a post-processing method for ASR error management which is independent of the ASR engine. We showed the error detection performance according to the voting threshold and achieved 0.514 of the F1-score best. We also evaluated the error correction method by word level error reduction rate by varying the detection voting threshold and achieved 8.21 % of the word error reduction rate when the voting threshold was 3. The error-corrected results is beneficial for spoken dialog applications by reducing the number of important erroneous words that cause unintended system operations.

Acknowledgements This work was supported by ICT R&D program of MSIP/IITP. [B0101-15-0307, Basic Software Research in Human-level Lifelong Machine Learning (Machine Learning Center).]

References

1. Brandow RL, Strzalkowski T (2000) Improving speech recognition through text-based linguistic post-processing, May 16 2000. US Patent 6,064,957
2. Jeong M, Jung S, Lee GG (2004) Speech recognition error correction using maximum entropy language model. In: Proceedings of INTERSPEECH. pp 2137–2140
3. Choi J, Lee D, Ryu S, Lee K, Kim K, Noh N, Lee GG (2014) Engine-independent asr error management for dialog systems. In: International workshop series on spoken dialogue systems technology (IWSDS)
4. Robinson T, Hochberg M, Renals S (1996) The use of recurrent neural networks in continuous speech recognition. In: Automatic speech and speaker recognition. Springer, New York, pp 233–258
5. Graves A, Mohamed A, Hinton G (2013) Speech recognition with deep recurrent neural networks. In: IEEE international conference on acoustics, speech and signal processing (ICASSP). pp 6645–6649
6. Mikolov T, Karafiat M, Burget L, Cernocký J, Khudanpur S (2010) Recurrent neural network based language model. In: INTERSPEECH. pp 1045–1048
7. Han D, Choi K (2007) A study on error correction using phoneme similarity in post-processing of speech recognition. In: The journal of the Korea institute of intelligent transport systems. The Korean institute of intelligent transport systems (Korean ITS). pp 77–86

A Protection Method of Mobile Sensitive Data and Applications Over Escrow Service

Su-Wan Park, Deok Gyu Lee and Jeong Nyeo Kim

Abstract Recently, mobile devices are gradually increasingly used even in companies and governmental institutions and the exchange of company data and military secret data through mobile devices is increased. As a result, the illegitimate leakage and collection of user data related to mobile devices has greatly increased. In this paper, we propose a method of managing the sensitive data of a mobile device using an escrow server. The proposed scheme is advantageous in that it is capable of storing the entrusted sensitive data of a user within the mobile device and verifying the validity of app software installed on the mobile device using an escrow server. Thereby it enhances data protection, minimizes damage resulting from the exposure of data attributable to the loss of a mobile device, and prevents the installation of illegitimate software.

Keywords Secure storage · Data protection · Sensitive data management

1 Introduction

As the spread of mobile devices having a communication function, such smartphones and tablet PCs, has suddenly increased, the illegitimate leakage and collection of user data related to mobile devices has also greatly increased. In particular, as mobile devices are gradually increasingly used even in companies and

S.-W. Park · J.N. Kim
Cyber Security Research Division, Electronics and Telecommunications
Research Institute, Daejeon, Korea
e-mail: parksw10@etri.re.kr

J.N. Kim
e-mail: jnkim@etri.re.kr

D.G. Lee (✉)
Department of Information Security, Seowon University, Cheongju, Korea
e-mail: deokgyulee@gmail.com

© Springer Science+Business Media Singapore 2016
J.J.(Jong Hyuk) Park et al. (eds.), *Advances in Parallel and Distributed Computing and Ubiquitous Services*, Lecture Notes in Electrical Engineering 368,
DOI 10.1007/978-981-10-0068-3_13

governmental institutions and the exchange of company data and military secret data through mobile devices is increased, there is growing concern about the potential leakage of data. Thus, mobile devices are increasingly becoming the target of hackers and cybercriminals.

Security concerns exist at all levels of the mobile ecosystem, including the mobile device hardware, the operating system layer and the mobile browser capabilities, however, nowhere is the threat of malware or malicious activity more challenging than at the application level. The sheer number of apps from different sources presents an enormous threat vector for malware and malicious activity, particularly as users are not always fully aware of how apps behave and are therefore lax in how they protect against potential unwanted behavior [1].

Mobile app stores provide access to billions of apps, some of which have been tested to some degree but many of which are only minimally vetted before being accepted for store distribution. For instance, while Apple is famous for its strict app vetting approach, Google has taken a more relaxed attitude and allows app developers greater freedom to upload apps to its store. Further, data leakage accidents in general occur because data is lost due to a user's carelessness or data leaks occur via software including malicious viruses or Trojan code, intentionally and unintentionally installed, through the accessing of the sensitive data of the user without a user's consent. Accordingly, there is a need for a method for management which is capable of enhancing the storage and protection of sensitive data in a mobile device.

Therefore, this paper proposes a method of managing the sensitive data of a mobile device using an escrow server. The rest of the paper is organized as follows. Section 2 describes about related works, and Sect. 3 presents a protection method of mobile sensitive data and applications over escrow service in detail. Lastly, we summarize and conclude in Sect. 4.

2 Related Works

The various security techniques and products for protecting data of mobile device have been recently introduced. Mobile data protection (MDP) products and services are typically defined as software security methods that enforce confidentiality policies by encrypting data, and then defending access to that encrypted data on the primary and secondary storage systems of end-user workstations. For providing MDP products, the mobile security client market has been nominated currently many players including familiar names Symantec, McAfee, Kaspersky, Trend Micro, Juniper, and Citrix, but no breakaway leader.

In the research field, software security analysis techniques share the goal of investigating smartphone apps for potential privacy violations [2–7]. PiOS [3] employs static data flow analysis techniques and is implemented for the Apple iOS system. The use of static analysis enables exploring broad execution paths including infeasible ones. However, it is prone to false positives because of well-known problems in static analysis such as alias and context sensitivity problems. Thereafter, dynamic information flow analysis techniques have proven useful for intrusion detection and malware analysis. BackTracker [4] keeps track of the causality of process-level events for backtracking intrusion. Panorama uses the instruction-level dynamic taint analysis for detecting information exfiltration by malware in Windows OS [6]. In addition, TaaS [2] proposed a service for automated software testing and CloudAV [5] proposed antivirus as an in-cloud network service in recent. However, utilizing these software security analysis techniques for app verification is both ineffective and inefficient since it is too complicated and dependent on the device type and version. Thus, we employ app verifier being offered as service for validation of smartphone apps over the cloud.

The proposed method relates generally to a method of managing the sensitive data of a mobile device and an escrow server for performing the method and, more particularly, to a method of managing the sensitive data of a mobile device and an escrow server for performing the method.

3 The Proposed Scheme

Figure 1 shows overview of environment in which the sensitive data of a mobile device is stored through an escrow server. The environment to which a method of storing the sensitive data of a mobile device is applied includes a mobile device of a user, an escrow server, and a user PC of the user.

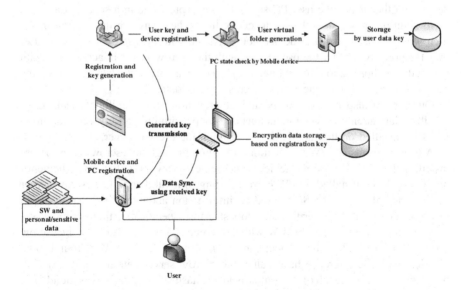

Fig. 1 Overview of an environment in which the sensitive data of a mobile device is stored through an escrow server

First, the procedure of registering the mobile device with the escrow server needs to be performed before the sensitive data of the mobile device is stored in the escrow server. The user requests the escrow server to register the mobile device via his or her mobile device. The escrow server performs the procedure of authenticating the mobile device and generates a user key EUK using information about the user and information about the mobile device. In this case, the generated user key, together with the information about the mobile device, is registered with the escrow server. At the same time, the generated user key is transmitted to the mobile device. Thereafter, the escrow server generates a user virtual folder configured to store the sensitive data and log information of the user, and establishes security storage policies so that the user virtual folder or the files of the user virtual folder are encrypted and stored. When the mobile device of the user is registered with the escrow server as described above, the sensitive data of the mobile device is encrypted, stored and managed in the user virtual folder within the escrow server according to user settings.

Furthermore, if the user requests the sensitive data of the mobile device to be shared by the user PC possessed by the user, the user may additionally perform the procedure of registering the user PC when registering the mobile device with the escrow server or after registering the mobile device with the escrow server. It may request data synchronization between the mobile device and the user PC from the escrow server. The sensitive data of the mobile device within the user PC synchronized by the escrow server may be encrypted and transferred by the mobile device or the escrow server. The sensitive data may be decrypted using the user key provided by the mobile device or the escrow server. In this case, the escrow server may check the state of the user PC while operating in conjunction with the mobile device. Further-more, the user PC stores the encrypted data in a specific database.

In addition, if the user desires to check his or her sensitive data stored in the escrow server through a user device not registered with the escrow server, the user may temporarily register the user device with the escrow server in accordance with a specific authentication procedure, may receive a one-time data key from the escrow server, and may check only corresponding data.

Our system also provides a method of preventing the illegitimate leakage of sensitive data attempted via a malicious app by performing validity verification on the S/W installed on the mobile device through the escrow server.

A user performs an S/W verification procedure through the escrow server before installing the S/W. The mobile device requests the escrow server to verify whether or not an app to be installed is S/W corresponding to a valid app. The escrow server verifies the validity of the S/W based on information about the S/W that has been received along with the verification request. More specifically, the escrow server compares the hash value of the S/W with a reference hash value stored in the escrow server. If, as a result of the comparison, the hash value of the S/W is found to be identical with the reference hash value, the escrow server determines the S/W to be legitimate. If, as a result of the comparison, the hash value of the S/W is found to be different from the reference hash value, the escrow server determines the S/W to be an illegitimate or modified file and transfers the results of the verification to the mobile device. The mobile device receives the results of the verification and installs the S/W only when the results of the verification indicate that the S/W is legitimate. In this case, the installation of illegitimate app S/W that is received through SMS, MMS or various messengers can be prevented.

The configuration of the mobile device is described as shown in Fig. 2. The mobile device includes a key management module, a sync management module, a storage management module, a data encryption/decryption module, and a software (S/W) verification request module.

The key management module manages a user key MEK generated by the mobile device and a user key EUK received from the escrow server. If the mobile device and the user PC desire to share sensitive data, the sync management module requests synchronization with the user PC from the escrow server, and establishes a synchronization policy corresponding to a request.

The storage management module determines whether sensitive data will be stored in the escrow server or the storage device of the mobile device, and determines whether or not to encrypt the sensitive data and establishes a storage policy based on the results of the determination. For this purpose, the storage management module includes a user mode storage module and a secret data storage module.

The data encryption/decryption module encrypts the sensitive data based on the results of the determination of the storage management module. For this purpose, the data encryption/decryption module includes a use encryption module configured to encrypt the sensitive data when the sensitive data is stored in the storage device of the mobile device, and an escrow encryption module configured to encrypt the sensitive data when the sensitive data is stored in the escrow server. The use encryption module encrypts the sensitive data using the user key MEK generated by the mobile device. The escrow encryption module encrypts the sensitive data using the user key EUK received from the escrow server. When the storage management module determines to store the sensitive data in the escrow server, the storage management module sends the sensitive data encrypted by the escrow encryption module to the escrow server. In contrast, when the storage management module

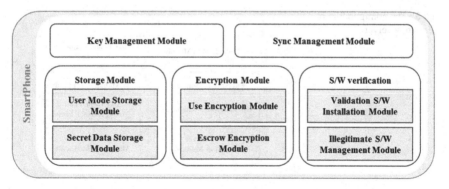

Fig. 2 The configuration of the mobile device

determines to store the sensitive data in the storage device of the mobile device, the user mode storage module and the secret data storage module operate in order to store the sensitive data in the storage device of the mobile device. The user mode storage module is a module configured to store common user data in a conventional storage device in a protected or non-protected state. The secret data storage module is a module configured to store sensitive data having high security strength in a secure depository that is logically or physically separated. This can be assumed that logically or physically secure depository has been provided in order to store sensitive data in a mobile device.

The S/W verification request module generates a hash value corresponding to information about S/W before the S/W is installed, encrypts the hash value using a user key EUK received from the escrow server, sends the encrypted hash value to the escrow server, and receives a result corresponding to the encrypted hash value. If the received result corresponds to valid S/W, the S/W verification request module installs the S/W using a validation S/W installation module. In contrast, if the received result corresponds to invalid S/W, the S/W verification request module stores information about the S/W and the hash value using the illegitimate S/W management module, and uses the information about the S/W and the hash value as S/W validation information within the mobile device itself.

The configuration of the escrow server is described as shown in Fig. 3. The escrow server includes a key management module, a virtual folder management module, a user data encryption/decryption module and an S/W validation module.

The key management module generates and manages the user key EUK of the registered mobile device. The virtual folder management module generates the user virtual folder of a registered device, that is, the mobile device, stores the sensitive data of the mobile device in the generated user virtual folder, and manages the user virtual in which the sensitive data is stored.

The user data encryption/decryption module re-encrypts encrypted sensitive data, received from the mobile device, according to a policy. For this purpose, the user data encryption/decryption module includes a user encryption module configured to perform encryption using the user key MEK generated by the mobile

Fig. 3 The configuration of the escrow server

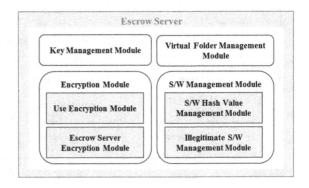

device, and an escrow server encryption module configured to perform encryption using the user key EUK generated by the escrow server.

The S/W validation module compares an S/W hash value received from the mobile device with the original hash value of an app managed by the S/W hash value management module, and transfers the results of the comparison to the mobile device. If, as the results of the comparison, the S/W hash value received from the mobile device is found to be different from the original hash value, the S/W validation module determines the corresponding S/W to be illegitimate and stores and manages the hash value and information about the S/W through the illegitimate S/W management module. The S/W hash value management module may refer to the original hash values of new apps registered with the escrow server for validation whenever the new apps of the mobile device are generated via a separate system, or may internally store and update information about the new apps.

As shown in Fig. 4, the user PC includes a key management module, a sync management module, a storage management module, and a data decryption module.

When a device is registered, the key management module manages a user key that is received from the escrow server and a key (e.g., a certificate) that may be received from the mobile device. The sync management module is set such that the sensitive data of the mobile device is synchronized in the user PC, and complies with the synchronization policy of the mobile device.

The storage management module obtains sensitive data from the escrow server in accordance with a user policy, and stores and manages the obtained sensitive data. In this case, the user PC may logically or physically separate its storage device like the mobile device. For this purpose, the storage management module includes a user mode storage module configured to store common data, and a secret data storage module configured to store sensitive data.

The data decryption module decrypts sensitive data stored in the storage management module. The data encryption/decryption module of the mobile device, the user data encryption/decryption module of the escrow server, and the data decryption module of the user PC are described in detail according to three types of cases below.

In the first case, the mobile device does not encrypt its sensitive data, but sends the sensitive data to the escrow server through a secure transport channel, and the

Fig. 4 The configuration of a user PC

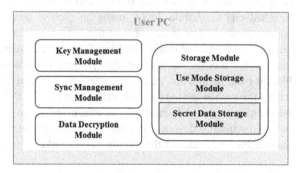

escrow server encrypts the sensitive data using a user key generated by the escrow server. Sensitive data within the user PC that has been synchronized may be decrypted and transmitted by the escrow server, or encrypted sensitive data may be transmitted so that it may be decrypted using a user key received from the escrow server when the device is registered. This method is advantageous in that data is securely stored in the escrow server, but is disadvantageous in that the data may be exposed on a transport channel and the escrow server may be aware of the original data.

In the second case, the mobile device encrypts its sensitive data using a user key EUK obtained from the escrow server when the device is registered. The sensitive data of the mobile device is secure because it is encrypted and transmitted. If the escrow server stores the encrypted data received from the mobile device without change, a performance load may be reduced because an encryption process is avoided. Furthermore, if the encrypted data received from the mobile device is further encrypted using the internal key of the escrow server and stored, high security may be guaranteed. The user PC that has been synchronized may easily decrypt data using a user key obtained from the escrow server, but there is a disadvantage in that the escrow server may be aware of the original data.

In the third case, the mobile device encrypts its sensitive data using a user key MEK generated by the mobile device. If the mobile device encrypts the sensitive data using its user key MEK and sends the encrypted sensitive data, a danger that the original data will be exposed may be avoided because the escrow server does not have the user key MEK. Furthermore, if the encrypted sensitive data is further encrypted using the internal key of the escrow server, security may be increased. However, it is cumbersome for the synchronized user PC to receive a key from the mobile device. This problem may be easily solved using the domain key of a user domain concept as described above.

4   Conclusion

In this paper, we propose a method of managing the sensitive data of a mobile device over an escrow server which is capable of storing the entrusted sensitive data of a user within the mobile device and verifying the validity of app software installed on the mobile device using an escrow server. The proposed method is advantageous in that it can protect the sensitive data of a user within the mobile device, can prevent damage resulting from the exposure of data attributable to the loss of the mobile device and can prevent the malicious exposure of data by blocking the installation of illegitimate S/W because the sensitive data of the user within the mobile device and the verification of S/W to be installed are managed via the escrow server. Advantageous in that it can protect the sensitive data of a user within the mobile device, can prevent damage resulting from the exposure of data attributable to the loss of the mobile device and can prevent the malicious exposure of data by blocking the installation of illegitimate S/W because the sensitive data of the user within the mobile device and the verification of S/W to be installed are managed via the escrow server.

Acknowledgments This work was supported by the ICT R&D program of MSIP/IITP. [R0101-15-0195(10043959), Development of EAL 4 level military fusion security solution for protecting against unauthorized accesses and ensuring a trusted execution environment in mobile devices.]

References

1. The Radicati Group, inc. "Mobile App Reputation Services", Understanding the role of App Reputation Services in delivering "Enterprise Grade" Mobile Security, A Whitepaper by The Radicati Group, Inc
2. Candea G, Bucur S, Zamfir C (2010) Automated software testing as a service. In: ACM SOCC, 2010
3. Egele M, Kruegel C, Kirda E, Vigna G (2011) "PiOS: detecting privacy leaks in ios applications", In: NDSS, 2011
4. King ST, Chen PM (2003) "Backtracking intrusions", In: SOSP, 2003
5. Oberheide J, Cooke E, Jahanian F (2008) "CloudAV: N-version antivirus in the network cloud", In: USENIX Security, 2008
6. Yin H, Song D, Egele M, Kruegel C, Kirda E (2007) "Panorama: capturing system-wide information flow for malware detection and analysis", In: ACM CCS, 2007
7. Poeplau S, Fratantonio Y, Bianchi A, Kruegel C, Vigna G (2014) Execute this Analyzing unsafe and malicious dynamic code loading in android applications. NDSS 14:23–26

GPU-Based Fast Refinements for High-Quality Color Volume Rendering

Byeonghun Lee, Koojoo Kwon and Byeong-Seok Shin

Abstract Color volume datasets of the human body, such as Visible Human or Visible Korean, describe realistic anatomical structures. However, imperfect segmentation of these color volume datasets, which are typically generated manually or semi-automatically, produces poor-quality rendering results. We propose an interactive high-quality visualization method using GPU-based refinements to support the study of anatomical structures. To smoothly represent the boundaries of a region-of-interest (ROI), we apply Gaussian filtering to the opacity values of the color volume. Morphological grayscale erosion operations are performed to shrink the boundaries, which are expanded by the Gaussian filtering. We implement these operations on GPUs for the sake of fast refinements. As a result, our method delivered high-quality result images with smooth boundaries providing considerably faster refinements, sufficient for interactive renderings as the ROI changes, compared to CPU-based method.

Keywords Color volume rendering · GPU-based refinement

B. Lee
Research and Development Department,
Zetta Imaging, Seoul, Republic of Korea
e-mail: intellee@gmail.com

K. Kwon · B.-S. Shin (✉)
Department of Computer Science and Information Engineering,
Inha University, Incheon, Republic of Korea
e-mail: bsshin@inha.ac.kr

K. Kwon
e-mail: mysofs@naver.com

© Springer Science+Business Media Singapore 2016
J.J.(Jong Hyuk) Park et al. (eds.), *Advances in Parallel and Distributed Computing and Ubiquitous Services*, Lecture Notes in Electrical Engineering 368,
DOI 10.1007/978-981-10-0068-3_14

1 Introduction

Direct volume rendering (DVR) efficiently visualizes medical images such as computerized tomography or magnetic resonance images [1]. Since these images do not contain optical information, DVR normally classifies voxels using a transfer function to map the voxel values to colors. Color volume datasets such as Visible Human (VH) [2] and Visible Korean (VK) [3] data provide anatomical structure information. However, assigning a region-of-interest (ROI) to color volume datasets using an opacity transfer function (OTF) is difficult due to the weak correlation between color values and organs. Hence, it is difficult to render color volume data using a general DVR method [4]. Color values are generally transformed into a monochrome space using various color spaces when color images are segmented [5]. OTF can be applied to the CIE L*U*V color space transformed from the RGB color space [6]. However, these methods are effective only for specific organs having a similar color distribution.

Segmentation data can be used as an additional volume data that contains ROI information. With this data, it is possible to render color volume datasets [7, 8]. However, segmentation is a time-consuming task because it is performed manually and these color volume datasets consist of thousands of slices. Furthermore, as illustrated in Fig. 1, the result may not be as precise due to complicated human anatomical structures. Since the segmentation data has a discrete volumetric form, it can cause aliasing artifacts while rendering.

There are two methods to render anatomical structures from the segmentation data: extracting surfaces [9, 10] and reconstructing a volumetric model [8, 11]. The use of a surface model to visualize the data by exploiting the existing graphics pipeline is straightforward, delivering smooth results and fast renderings. However, it is difficult to observe the interior of objects, which is highly demanded for anatomical structures. DVR can represent the interior of objects using a volumetric model. However, the rendering quality is dependent on the resolution of the volume.

(a)   **(b)**   **(c)**

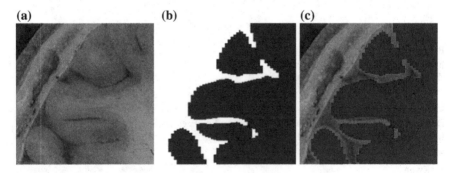

Fig. 1 Example of segmentation error. **a** is a part of the brain, **b** is the segmentation result of (**a**) and **c** is the overlaid image using (**a**) and (**b**)

We propose an interactive high-quality color volume rendering method with a volumetric model using fast refinements of the segmentation data. Opacity values of the volume data are filled using the segmentation data and user-defined indexed opacity table. To remove artifacts on the boundaries caused by discrete segmentation, we apply Gaussian filtering to the opacity values of the volume data. A morphological grayscale erosion operation is performed to shrink the boundaries that are over-expanded due to the Gaussian filtering. Since these operations can be highly parallelized, we implement these operations on graphical processing units (GPUs) for the sake of fast refinements. Our method provides realistic renderings using a real color volume.

2 Description of Method

2.1 Boundary Refinements

We refine the boundaries of the opacity values of the color volume data. Gaussian filtering is applied to the opacity values of the volume data. Then, morphological grayscale erosion is performed to shrink the boundaries that are over-expanded by the Gaussian filtering.

Generally, color human body data contains RGB values only. Opacity values are assigned to the alpha channel of the volume data by using segmentation data and user-defined indexed opacity table [11]. Since segmentation tasks are performed on a discrete image space, these values are formed discretely resulting in artifacts while rendering. We apply Gaussian filtering to the opacity values of the volume data to smooth the object boundaries. Our method can apply the Gaussian filter with different r and σ for each axis to reduce discontinuities for datasets having a larger spacing between the slices than the pixel spacing. This is described as follows:

$$f(x_0, y_0, z_0) = \int_{-\infty}^{\infty} \int_{-\infty}^{\infty} \int_{-\infty}^{\infty} g(x, y, z)I(x_0 - x, y_0 - y, z_0 - z)dxdydz, \quad (1)$$

where (x_0, y_0, z_0) is the position to be filtered, $g(x, y, z)$ is the Gaussian weight. After the Gaussian filtering, the ROI of the original object is expanded according to the radius of the filter. This may visualize undesirable parts of adjacent objects. Furthermore, there are unwanted mixtures, which need to be peeled out. To remedy this problem, we apply morphological grayscale erosion operations into the opacity values for each axial slice after the Gaussian filtering. A disk shaped structuring element, which means D_b is a circle centered on s in Eq. 2, is used in this paper. The radius of the structuring element, which determines the extent of D_b, is dependent on the pixel pitch of the volume data and the radius of the Gaussian filter.

$$(f \ominus b)(s) = \min\{f(s+x) - b(x)|(s+x) \in D_f, x \in D_b\}. \quad (2)$$

2.2 Fast GPU-Based Refinement

Applying Gaussian filtering to the opacity values of the volume data requires significant computation. This could prevent our method from providing interactive renderings when the ROI changes because the opacity values of the new ROI must be filtered. To achieve interactive filtering as the ROI changes, we implement GPU-based Gaussian filtering. Since Gaussian filtering can be performed in highly parallel, GPUs can accelerate it remarkably. For a further speed improvement, we use *separable filters* that have the time complexity of $O(nr)$ where n and r are the volume data size and the radius of the filter, respectively, whereas the complexity of the brute-force method is $O(nr^3)$.

We further optimize the filtering by exploiting single instruction multiple data (SIMD) operations that GPUs natively support. GPUs can process a vector containing four scalar values, such as an RGBA channel. This can provide an additional speed-up because the number of instructions and threads is reduced. Our method filters four grayscale slices simultaneously by packing them into one RGBA slice.

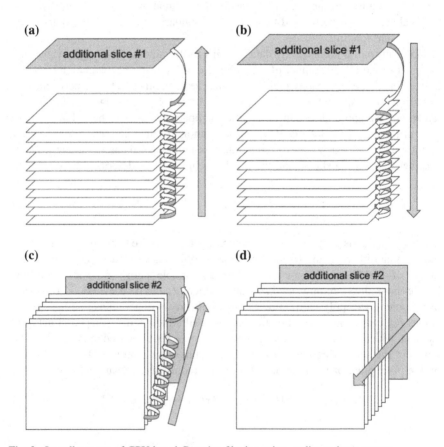

Fig. 2 Overall process of GPU-based Gaussian filtering using cyclic render-to-texture

We also minimize the memory overhead by using the manner of cyclic render-to-texture [12]. Figure 2 presents the overall procedure for filtering the opacity values. We use only two additional slices, which have sizes of $L_x \times L_y$ and $L_x \times L_z$, where L_x, L_y, and L_z are the sizes of each axis of the volume data. At the x-direction filtering stage, each filtering result is stored in the upper slice of the current slice (Fig. 2a). The result of the top slice is stored in an $L_x \times L_y$-sized additional slice as an exception. The storing direction is the opposite of the previous stage at the y-direction filtering stage (Fig. 2b). The z-direction filtering stage is performed with the other additional slice similar to the x-direction stage (Fig. 2c). The filtering is completed after the *slice-shift-back* step to store the result into the original volume space (Fig. 2d). This minimization of the memory overhead is advantageous because there can be a memory space limitation on the GPU, especially for a large dataset.

The erosion operation can be performed on the GPU intuitively [13]. We further optimize this by exploiting SIMD operations.

3 Results and Discussion

Our method is implemented in Direct3D 11 and HLSL on an NVIDIA GTX 780 Ti with 3.0 GB of memory. All experiments were performed on an Intel i7 CPU with 16 GB RAM and the GPU described above. *Early ray termination* and *empty space leaping* were applied for rendering speed optimization. The rendering image resolution is 800×800 in all cases.

We conducted experiments using the torso dataset ($494 \times 282 \times 405$), which contains many organs, to demonstrate the effectiveness of our method. Figure 3 shows the rendering results with various refinements of the segmentation data. Figure 3a is the result without any refinement. As shown, the surface was rendered roughly because the segmentation data is represented as a discrete form. We applied a 3D Gaussian filter with $r = 3$ and $\sigma = 1$ (Fig. 3c). Consequently, the surface was rendered noticeably smoother than Fig. 3a. However, some artifacts were observed at the boundaries of multiple objects meeting each other (yellow circle on Fig. 3e). The rightmost column of Fig. 3 shows the result images that were applied by erosion after the Gaussian filtering. As shown, the artifacts between the ribs and lungs were removed (yellow circle on Fig. 3f). This shows that our refinements are effective for rendering smooth boundaries.

Table 1 shows the refinement times of Fig. 3, which are averaged from ten experimental executions. As shown, our basic GPU-based method was approximately 46 times faster than the CPU-based method with O(nr) time complexity. Our GPU-based optimization using SIMD gave about 86 speed-ups in comparison to the CPU-based method, and consequently, interactive renderings as the ROI changed became feasible.

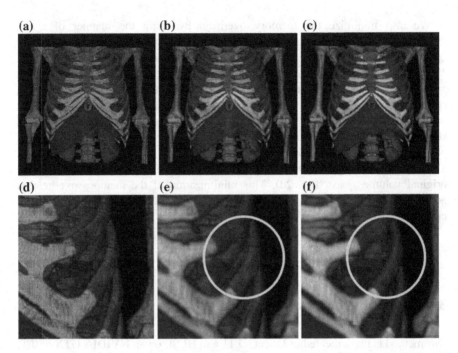

Fig. 3 Rendering results using various refinements: **a** nothing, **b** Gaussian filtering, and **c** Gaussian filtering and erosion. **d, e, f** are magnified images of (**a**), (**b**), and (**c**), respectively

Table 1 Volume refinement time using the torso dataset (unit: ms)

Device	Gaussian filtering ($r = 8$)	Erosion ($r = 5$)	Total
CPU	1530	4931	6461
GPU	48.75	92.5	141.25
GPU (SIMD opt.)	23.75	51.25	75
Speed-up	64.42	96.21	86.15

4 Conclusion

We proposed an interactive high-quality color volume rendering method that employs Gaussian filtering, morphological grayscale erosion. Our method applied a Gaussian filtering to the opacity values of the color volume data resulting in smooth boundaries. The morphological grayscale erosion operation effectively contracted the ROI that was expanded due to the Gaussian filtering. Moreover, erosion removed the artifacts caused by multiple objects facing each other. Our GPU-based refinement was approximately 86 times faster than the CPU-based method. This allowed interactive rendering whenever the ROI changed. By implementing a cyclic manner, our approach also minimized the memory overhead caused by refining.

Acknowledgments This work was supported by the National Research Foundation of Korea (NRF) grant funded by the Korea government (MSIP) (No. 2015R1A2A2A01008248). This research was supported by Next-Generation Information Computing Development Program through the National Research Foundation of Korea (NRF) funded by the Ministry of Science, ICT & Future Planning (2012M3C4A7032781).

References

1. Levoy M (1988) Display of surfaces from volume data. IEEE Comput Graphics Appl 8(3):29–37
2. Ackerman MJ (1998) The visible human project. Proc IEEE 86(3):504–511
3. Park JS, Chung MS, Hwang SB, Lee YS, Har D-H, Park HS (2005) Visible Korean human: improved serially sectioned images of the entire body. IEEE Trans Med Imaging 24(3):352–360
4. Liu Y, Chen JX, Yang L (2008) Real-time photorealistic virtual human anatomy. Comput Sci Eng 10(2):41–47
5. Pal NR, Pal SK (1993) A review on image segmentation techniques. Pattern Recogn 26(9):1277–1294
6. Ebert DS, Morris CJ, Rheingans P, Yoo TS (2002) Designing effective transfer functions for volume rendering from photographic volumes. IEEE Trans Visual Comput Graph 8(2):183–197
7. Park JS, Chung MS, Hwang SB, Lee YS, Har D-H (2005) Technical report on semiautomatic segmentation using the adobe photoshop. J Digit Imaging 18(4):333–343
8. Kwon K-J, Shin B-S (2006) Visualization of segmented color volume data using GPU. In: Advances in artificial reality and tele-existence, Springer, 1062–1069
9. Shin DS, Park JS, Park HS, Hwang SB, Chung MS (2012) Outlining of the detailed structures in sectioned images from visible Korean. Surg Radiol Anat 34(3):235–247
10. Shin DS, Chung MS, Lee JW, Park JS, Chung J, Lee S-B, Lee S-H (2009) Advanced surface reconstruction technique to build detailed surface models of the liver and neighboring structures from the visible Korean human. J. Korean Med Sci 24(3):375–383
11. Uhl J-F, Park JS, Chung MS, Delmas V (2006) Three-dimensional reconstruction of urogenital tract from visible Korean human. Anat. Rec. Part A: Discoveries Mol Cell Evol Biol 288(8):893–899
12. Yan G, Tian J, Zhu S, Dai Y, Qin C (2008) Fast cone-beam CT image reconstruction using GPU hardware. J X-ray Sci. Technology 16(4):225–234
13. Rane MA (2013) Fast morphological image processing on GPU using CUDA. Ph.D. thesis. College of Engineering, Pune

Beacon Distance Measurement Method in Indoor Ubiquitous Computing Environment

Yunsick Sung, Jeonghoon Kwak, Young-Sik Jeong
and Jong Hyuk Park

Abstract In the indoor ubiquitous computing environment where Global Positioning System (GPS) cannot be utilized, the approach to calculate the locations of Unmanned Aerial Vehicles (UAVs) is the core technique to control multiple UAVs. To calculate the locations of UAVs, the distance between Access Points (APs) and UAVs should be measured accurately given that the location of UAVs is obtained on the basis of the distance between APs and UAVs. In this paper, we propose a method to measure the distance between a single beacon and a single AP in an indoor ubiquitous computing environment. We assume that the beacon is attached to the bottom of a UAV. In the indoor experiment, while transferring a beacon, the distances between the beacon and an AP were measured and tuned. Therefore, the accumulated difference between the real beacon location and the calculated beacon location was reduced by 31.1 %.

Y. Sung (✉)
Faculty of Computer Engineering, Keimyung University, Daegu, South Korea
e-mail: yunsick@kmu.ac.kr

J. Kwak
Department of Computer Engineering, Graduate School, Keimyung University,
Daegu, South Korea
e-mail: jeonghoon@kmu.ac.kr

Y.-S. Jeong
Department of Multimedia Engineering, Dongguk University, Seoul, South Korea
e-mail: ysjeong@dongguk.edu

J.H. Park
Department of Computer Engineering, Seoul National University
of Science & Technology, Seoul, South Korea
e-mail: jhpark1@seoultech.ac.kr

© Springer Science+Business Media Singapore 2016
J.J.(Jong Hyuk) Park et al. (eds.), *Advances in Parallel and Distributed Computing and Ubiquitous Services*, Lecture Notes in Electrical Engineering 368,
DOI 10.1007/978-981-10-0068-3_15

1 Introduction

Recently, as diverse types of Unmanned Aerial Vehicles (UAVs) have been released, the number of different types of autonomous UAV-based services has increased [1–4]. In the field of UAVs, Global Positioning System (GPS) is usually utilized to measure the locations of UAVs and to control the UAVs. However, from the viewpoint that GPS cannot be utilized in an indoor environment, a novel approach is required to estimate the locations of UAVs.

Beacons can be applied to measure the locations of UAVs in an indoor environment [5–7]. Each beacon can be attached to an UAV and then be utilized to obtain the locations of the UAV. The core issue of obtaining the locations of UAVs is the accuracy of measuring the distances between Access Points (APs) and beacons where the APs receive the Bluetooth signals of beacons.

In this paper, we propose a beacon distance measurement method for an indoor ubiquitous computing environment. The beacon is attached to the bottom of an UAV, and the locations of the UAV are estimated by utilizing an AP, which can measure their distances to the beacon.

The rest of this paper is organized as follows: Sect. 2 proposes a beacon distance measurement method. Section 3 discusses the validation of the proposed method. Finally, Sect. 4 concludes this paper.

2 Filter-Based Beacon Distance Measurement

2.1 Beacon Distance Measurement Framework

In the given indoor environment, each beacon is attached to the bottom of the corresponding UAV. The Bluetooth signal of each beacon is transferred to all the reachable APs. The APs recalculate the distance between all beacons and themselves and then, transfer their results to a ground station. Figure 1 shows a beacon-based UAV-ground station framework.

Each AP has two modules and one pool: Bluetooth module, filter pool, and Wi-Fi module. The Bluetooth module receives the signals of beacons. The filter pool contains filters that handle and recalculate the received signals, such as beacon signals and the recalculated distances. The Wi-Fi module sends the results of the filter pool to a ground station.

2.2 Definition of Indoor Environment

In the UAV environment, the locations of points are predefined and fixed to set the path where UAVs need to be flown according to the path as shown in Fig. 2. Therefore, the location information of the points and the distance between the

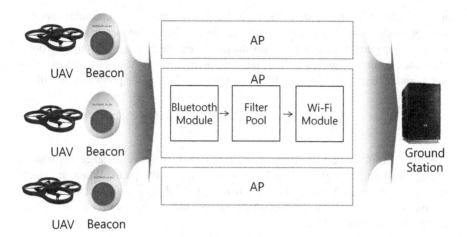

Fig. 1 Beacon-based UAV-ground station framework

Fig. 2 Beacon-based
UAV-ground station
framework

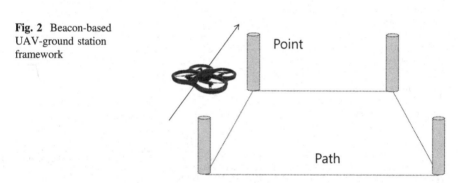

points and APs can be utilized to revise the distance between beacons and the APs. For the description of the proposed method clearly, we assume that there is a single AP and a single UAV.

A UAV flies from a predefined starting point to one of other points according to the predefined order of points, which denotes the path of the UAV. We assume that an UAV flies from one point to another in a straight line without moving in a zigzag pattern. When the UAV reaches a specific point, it stops flying for a while and then starts to fly to the next point. By stopping its flight, the UAV has a chance to adjust its location on the basis of a corresponding beacon and an AP.

2.3 Filters of Beacon Distances

In the proposed method, the following filters are suggested. First, the point location filter utilizes the location information of points when a beacon starts to be transferred from a specific point and adjusts the distance from the beacon. The distance of the beacon is recalculated, as shown in Eq. (1). α denotes the minimum distance between a point and an AP, and β represents the maximum distance between a point and an AP. φγ denotes the top γ % distances among the measured distances from a last visited point. d_t represents the distance between a last visited point and an AP at time t, and d'_t denotes the recalculated distance at time t by the point location filter.

$$d'_t = (d_t - \varphi_\gamma) \times \frac{\beta - \alpha}{\varphi_{100-\gamma} - \varphi_\gamma} + \alpha. \tag{1}$$

Second, a hovering filter assigns the distance d_t with the distance between a current point and an AP, while a UAV is hovering at the point. When the UAV starts to fly, the point location filter is utilized, again. Although this paper proposes two types of filters, diverse types of filters can be added to the filter pool considering the environment and UAVs.

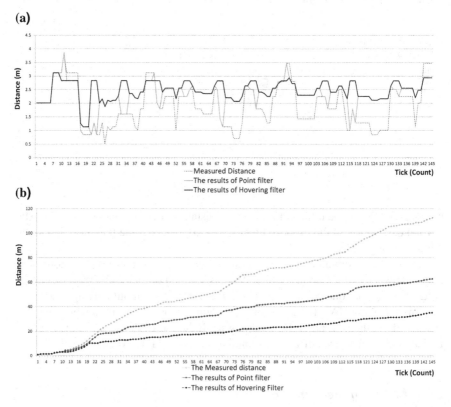

Fig. 3 Distances between an AP and a beacon while transferring the beacon. **a** Measured distances and filtered distances **b** Difference between planned distances and the distances measured and filtered

3 Experiment

In an indoor experiment, one beacon and one AP were utilized. After setting the AP at the center of the indoor environment, four points were defined as the locations that define the rectangular path of the UAVs around the AP. While the beacon was moved around the AP reaching each point in order, the distances between the AP and the beacon were measured. Given that the beacon was moving when the distances were measured and the experimental environment was inside a building, there was a gap between the real distances and the measured distances between a beacon and an AP.

In this experiment, a point location filter and a hovering filter were applied sequentially. α, β, and γ were set to 2, 2.8, and 0, respectively. Figure 3 shows the measured and calculated distances according to the transfer of the beacon by utilizing two filters. The measured distances range from 49.5 to 386.0 cm. The point location filter adjusts the measured distances first. By applying the hovering filter, the distance when the beacon reached a specific point was changed to 28.2 cm. As a result, the accumulated difference from an AP and a beacon was reduced from 112,485 to 35,000 cm.

4 Conclusion

This paper proposed a distance measurement method between a single beacon and a single AP. By increasing the accuracy of the distance measurement, the accuracy of the locations of UAVs can be increased given that the locations of UAVs were calculated on the basis of the distance between the beacons and the APs.

In the experiment, two types of filters, namely point location and hovering filters, were utilized to adjust the beacon distances. The original distance from a beacon had critical errors and hence, could not be utilized directly by a ground station. However, by applying the proposed method, the distance between a beacon and an AP could be accurately adjusted. By applying the proposed method, the accumulated difference between the real beacon location and the calculated beacon location was reduced by 31.1 %.

Acknowledgements This research was supported by Basic Science Research Program through the National Research Foundation of Korea (NRF) funded by the Ministry of Science, ICT & Future Planning (NRF-2014R1A1A1005955).

References

1. Nitschke C, Minami Y, Hiromoto M, Ohshima H, Sato T (2014) A quadrocopter automatic control contest as an example of interdisciplinary design education. In: Proceeding of 14th international conference on control, automation and systems. KINTEX, Gyeonggi-do, Korea, pp 678–685
2. Spek JVD, Voorsluys M (2012) AR.Drone autonomous control and position determination. Bachelor thesis, Delft University of Technology

3. Dijkshoorn N (2012) Simultaneous localization and mapping with the AR.Drone. Master thesis, Universiteit Van Amsterdam
4. Bristeau P, Callou F, Vissière D, Petit N (2011) The navigation and control technology inside the AR.Drone micro UAV. In: Proceeding of international federation of automatic control, vol 18, no 1. Milano, Italy, pp 1477–1484
5. Lee HC, Lee DM (2011) A Study on localization system using 3D triangulation algorithm based on dynamic allocation of beacon node. J Korea Inf Commun Soc 36(4):378–385
6. Lee HC, Lee DM (2013) The 3-Dimensional localization system based on beacon expansion and coordinate-space disassembly. J Korea Inf Commun Soc 38B(1):80–86
7. Liu H, Darabi H, Banerjee P, Liu J (2007) Survey of wireless indoor positioning techniques and systems. IEEE Trans Syst Man Cybern Part C Appl Rev 37(6):1067–1080

Indoor Location-Based Natural User Interface for Ubiquitous Computing Environment

Jeonghoon Kwak and Yunsick Sung

Abstract The locations of residents are utilized not only for recognizing the situation of indoor ubiquitous computing environments but also for controlling devices and expressing intention. This paper proposes a novel user interface to utilize the locations of users as control signals in indoor ubiquitous computing environments. The locations of users are estimated by analyzing the distance between a beacon and APs.

Keywords NUI/NUX · Human-computer interaction · Indoor location

1 Introduction

Knowledge of residents' locations in indoor ubiquitous computing environments is the key to understand or estimate their activities. There are diverse kinds of approaches to recognize and utilize the locations of residents.

Estimated locations can be utilized as control signals by applying the concepts of natural user interface/natural user experience (NUI/NUX) for user-friendly control interfaces. For example, one research study utilized Kinect, a multiple-depth camera, to recognize the locations of people [1]. Beacon is also utilized to calculate locations in indoor environments [2–4]. The depths from each camera are integrated into a 3D model of a person. Leap Motion is utilized to recognize hand locations and improve the problem of recognizing sign language [5].

J. Kwak
Department of Computer Engineering, Graduate School,
Keimyung University, Daegu 42601, South Korea
e-mail: jeonghoon@kmu.ac.kr

Y. Sung (✉)
Faculty of Computer Engineering, Keimyung University,
Daegu 42601, South Korea
e-mail: yunsick@kmu.ac.kr

© Springer Science+Business Media Singapore 2016
J.J.(Jong Hyuk) Park et al. (eds.), *Advances in Parallel and Distributed Computing and Ubiquitous Services*, Lecture Notes in Electrical Engineering 368,
DOI 10.1007/978-981-10-0068-3_16

This paper proposes an NUI/NUX that estimates the locations of users in indoor ubiquitous computing environments. The locations of users are calculated based on the inaccurate distances between beacons and APs and then utilized as control signals. In the experiments, four APs were utilized to measure the distance from a beacon, and the four distance measurements were utilized to estimate the location of a user.

The rest of this paper is organized as follows: Sect. 2 proposes a location-based NUI/NUX interface. Section 3 discusses the validation of the proposed method. Finally, Sect. 4 concludes this paper.

2 User Location Estimation Framework in Indoor Environment

2.1 Location-Based NUI/NUX Concept

In indoor environments, APs and beacons are utilized to estimate the locations of users. After deploying APs and providing each user with one beacon, the APs begin to measure the distances from the beacons and then calculate the locations of users based on the measured distances.

As shown in Fig. 1, the locations of users are estimated with respect to the locations of APs. When users move around APs, their locations are denoted as spaces relative to the center among four APs—i.e., front space s_f, down space s_b, left space s_l, and right space s_r.

The relative spaces of the users on the basis of the center are utilized as control signals. However, given that Bluetooth signals are usually blocked by a user's body, it is better to deploy APs on the ceiling.

Fig. 1 Locations of users with beacons and APs

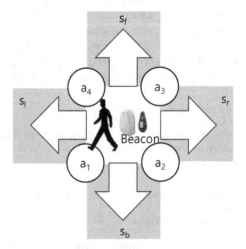

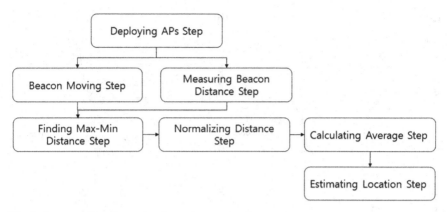

Fig. 2 Process of location estimation

2.2 Process of User Location Estimation

This paper defines five steps between initialization process and estimation process to obtain the distances between APs and a beacon that are utilized for the estimation of user locations as shown in Fig. 2. During the initialization process, APs are deployed in the environment where the movement of users is to be observed. By deploying APs, the four parameters—front space s_f, back space s_b, left space s_l, and right space s_r—are defined.

First, a user moves with a beacon during the space visit step. To measure the distances from the beacon in terms of the four spaces, the user must visit the four spaces. Second, while the user is moving, the four distances between the beacon and the APs are measured. Third, after the user has moved to obtain the four distances, the maximum and minimum distances among the obtained measurements are obtained. Fourth, all obtained measurements are normalized from 0 to 1 relative to the maximum and minimum distances. Fifth, the average of the normalized distances is calculated. All steps are for obtaining the information necessary to estimate the locations of users. After the fifth step, the users can move freely, and their locations are estimated based on the maximum distance, minimum distance, and average of the normalized distances.

3 Experiment

In the experiment, four APs were deployed and one beacon were utilized with 1 m maintained between each AP. A user visited each space two times in order from AP a_1 to AP a_4. Figure 3 shows the distance values of the beacon while a user stayed at each space. Figure 4 shows the normalized measurements after determining the

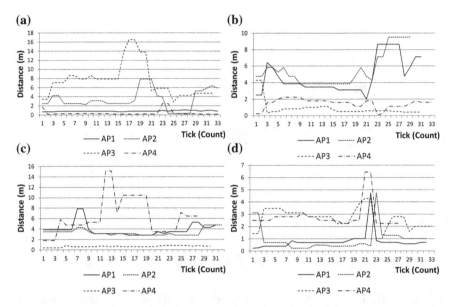

Fig. 3 Distances in four spaces. **a** Distances in left space s_l, **b** distances in front space s_f, **c** distances in right space s_r, **d** distances in back space s_d

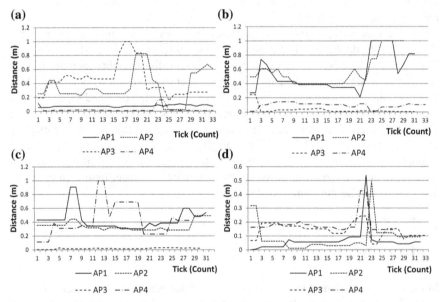

Fig. 4 Normalized location distances. **a** Left location normalized distance, **b** front location normalized distance, **c** right location normalized distance, **d** back location normalized distance

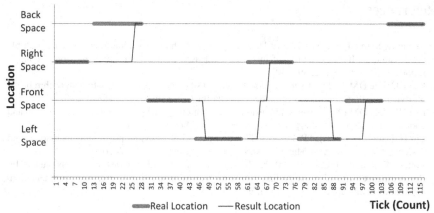

Fig. 5 Results of user locations

maximum and minimum distances. It was confirmed that the normalized distances were between 0 and 1.

Figure 5 shows the results of the recognized spaces. The measurements of a user outside the four spaces are not denoted. When the user was inside the four spaces, the space recognition rate was 63.3 %. However, at the end of each stay at a given space, the proposed method correctly recognized the space at which the user was located.

4 Conclusion

This paper proposed a space estimation-based NUI/NUX for indoor ubiquitous computing environments. Given that the accuracy of the distances from beacons to APs in indoor environments is too low to be utilized, the proposed method is an approach that can be utilized in such an environment.

The proposed method can be utilized in a diversity of applications such as indoor game interfaces and residence life analysis. Given that NUI/NUX is familiar to residents, products based on NUI/NUX can be implemented without repulsion.

Acknowledgements This research was supported by the MSIP(Ministry of Science, ICT and Future Planning), Korea, under the ITRC(Information Technology Research Center) support program (IITP-2015-H8501-15-1014) supervised by the IITP(Institute for Information and communications Technology Promotion).

References

1. Zhang L, Sturm J, Cremers D, Lee D (2012) Real-time human motion tracking using multiple depth cameras. In: Proceedings of the 2012 IEEE/RSJ international conference on intelligent robots and systems, pp 2389–2395
2. Lee HC, Lee DM (2011) A study on localization system using 3D triangulation algorithm based on dynamic allocation of beacon node. J Korea Inf Commun Soc 36(4):378–385
3. Lee HC, Lee DM (2013) The 3-dimensional localization system based on beacon expansion and coordinate-space disassembly. J Korea Inf Commun Soc 38B(1):80–86
4. Liu H, Darabi H, Banerjee P, Liu J (2007) Survey of wireless indoor positioning techniques and systems. IEEE Trans Syst Man Cybern Part C Appl Rev 37(6):1067–1080
5. Potter LE, Araullo J, Carter L (2013) The leap motion controller: a view on sign language. In: Proceedings of the 25th Australian computer-human interaction conference: augmentation, application, innovation, collaboration, pp 175–178

Flexible Multi-level Regression Model for Prediction of Pedestrian Abnormal Behavior

Yu-Jin Jung and Yong-Ik Yoon

Abstract The high incidence of heinous crime is increasing to use of CCTV. However, CCTV has been used to obtain evidence rather than crime prevention. Also it shows a weak effect about preventing crime. To solve the weak effort, we propose a Flexible Multi-level Regression (FMR) model that should estimate a dangerous situation for the pedestrian. The FMR model is tracking the behavior of between pedestrians from multiple CCTV that are located in different locations. The FMR has a prediction logic that should estimate an abnormal situation to analyze the possibility of crime by using the Regression and Apriori algorithm. The FMR model can be usefully used to prevent the crime because of an immediate response and rapid situation assessment.

Keywords CCTV systems · Flexible multi-level regression · Behavior prediction · Abnormal behavior · Situation assessment

1 Introduction

The people at daily crime news are feeling social anxiety. Society and people in the occurrence of so-called hate crimes has been heightened anxiety about security including the five major crimes, such as murder, violence, robber, rape, theft [1, 6]. Accordingly, the recognized about security is also increasing. However, the concept of hate crime has problems can be slightly ambiguous that property about behavior does not appear. The people are concerned about how to prevent crimes before they occurs a situation. In addition, most populated area is gradually installing CCTV. The CCTV of an area as an extension the related technologies also have developed

Y.-J. Jung · Y.-I. Yoon (✉)
Department of Multimedia Science, SookMyung Women's University, Seoul, Korea
e-mail: yiyoon@sm.ac.kr

Y.-J. Jung
e-mail: sdj4351@naver.com

© Springer Science+Business Media Singapore 2016   137
J.J.(Jong Hyuk) Park et al. (eds.), *Advances in Parallel and Distributed Computing and Ubiquitous Services*, Lecture Notes in Electrical Engineering 368,
DOI 10.1007/978-981-10-0068-3_17

constantly [2, 8]. However, the rate of growth and development is evolving rapidly, but the maturity of the technology is still a lack [3].

Therefore, in this study, we propose a prediction model for abnormal behavior pattern by behavior and environment information collected from the CCTV. In addition, the Flexible Multi-level Regression (FMR) model will establish that through environmental information can estimate risk of behavior and abnormal behavior. It can be seen the risk of abnormal behavior and the behavior of object. Then, the final step is to predict the behavior through the associated rules of behavior. Apriori algorithm using behavior data collected from the CCTV are to predict the next move.

2 Flexible Multi-level Regression Model for Prediction of Abnormal Behavior

In this section, we first describes classify data. The classified data is applied to the Flexible Multi-level Regression (FMR) model. It will be described abnormal behavior diagram of prediction system that can predictable for abnormal behavior [7]. Situation awareness and prediction part is expected to describe [4]. Mainly we focus on situation awareness. At first, we define abnormal behaviors in this study. The abnormal behaviors are recognized when the behaviors from a general view to normal movement.

2.1 Abnormal Behavior Diagram of Prediction System

Figure 1 shows a configuration diagram for predicting the abnormal behaviors. Context awareness is the step of recognizing and tracking objects. At first, dynamic objects are determined by detecting a movement of objects from the CCTV. Behavior of objects information is filtered from collected objects information. The area information and the environment information are collected, including information of the object. The behaviors of objects are watching and identify the distance between objects. At this time, it puts the weights by analyzing the position information and environmental information [5, 6]. It passes the analysis of the behavior of the object information to the next step. This is a step to capture the abnormal behavior by behavior analysis of the passed object. The FMR model can used to analyze the collected environment information and the behavior of object. Then, it is determined whether or not the abnormal behavior of the object. Through the association rules of behavior in the last step collects data from the behavior using the association rule of Apriori algorithms, it becomes possible to predict the actions to execute.

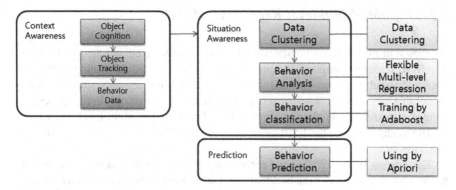

Fig. 1 Configuration of behavior prediction

2.2 Data Classification

The data selected variables on abnormal behavior according to the criminal psychology and sociology. This data is a variable influencing the abnormal behavior from the CCTV. In this paper, the input data consists of the distance, area, time, and the weather.

- Distance information

Information refers to a measure of the distance between the recognized objects with the image. And it includes identification information about the object. However, it is assumed that objects are independent of each other. The distance between the objects classifies four stages by proxemics.

- Area information

Area includes information about the CCTV location or path. It is mainly used as the information for determining the risk area. The analysis area is mainly the width of the road.

- Time information

The present time is identified in CCTV. Time is largely divided into a morning/morning/morning/afternoon/evening.

- Weather information

Weather includes weather information such as the amount of clouds and brightness. The risk level is considered by the amount of cloud and brightness. In this study, we set the amount of clouds with weather information.

2.3 Flexible Multi-level Regression Model

As shown in Fig. 2, the behavior pattern is classified to fixed and flexible features. The fixed features consist of distance and area and the flexible features consist of time and weather. These features are used in regression analysis step to build a model that can estimate whether a risk of behavior occurs or not. In this paper, we defined a model of combinational analysis, called Flexible Multi-level Regression (FMR) model. Finally, the FMR generates the result that is predicted the abnormal behavior.

The process of the FMR is red color module as shown schematically in Fig. 2. Set the distance and area variables that affect the abnormal behavior in fixed feature, the variable of the time variable and the weather was set to flexible feature. Weather variables usually are processed missing correlation coefficient when there is no change, not analyzed, when the weather has changed and treated as a variable that is flexible in the variable to be analyzed. It is analyzed by the fixed analysis when they are treated as missing values, if there are no missing values derives both results through again the combination analyzes analysis. In this paper, we describe only fixed feature, it was constructed a regression model such as Eq. (1).

$$\text{Behavior} = 5.603 + 0.517\,(\text{distance}) + 0.222\,(\text{area}). \tag{1}$$

Equation (1) are possible to estimate the behavior. It indicates numerical value of the distance is such that changes in how the behavior when the increased one unit. In other words, distance is 1 increase, has a meaning that would like to increase by 0.517 units. Therefore, the behavior of when the distance and area increased by 1 becomes 6.342. The results came out assessment of behavior is performed as shown in Table 1. The result value is to come out risk corresponding to each range, it is possible to estimate whether the hazard behavior.

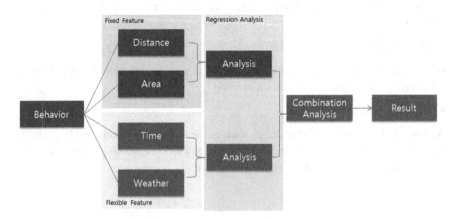

Fig. 2 Flow diagram of FMR

Table 1 Assessment index of behavior

Range	Degree of risk
~23	Dangerous
24–55	Caution
56~	Normal

Table 2 Implementation result of the Apriori

	lhs		rhs	Support	Confidence	Lift
1	{standing}	=>	{running}	0.375	0.750	1.000
2	{running}	=>	{standing}	0.375	0.500	1.000
3 .	{punching}	=>	{running}	0.500	1.000	1.333
4	{running}	=>	{punching}	0.500	0.667	1.333

2.4 Behavior Prediction

In this section, it predicts the behavior through the analysis of behavior. Behavior prediction uses a representative of the association rules of Apriori algorithm. Association rules help in finding the association between the data and is also used to find the relationship between the attributes.

Table 2 shows an implementation of the association data by applying Apriori algorithm at the behavior data. The properties of this result are the attributes relevant to each other. For example, in the case of the 4th, the behavior of the object 1 to track when the "running" means that the following action is high if the "punching". With this result, it is possible to derive the association rules for the actions and behavior of dangerous objects by analyzing the behavior, and then to know in advance.

3 Scenario Application

Figure 3 shows the estimated scenario Flexible Multi-level Regression Model for abnormal behavior prediction. In the first (a) $Area_1$ as shown in Fig. 3, $Camara_1$ determines the person's dynamic objects by sensing. The person's objects are recognized after grasping the distance between behavior and the area for the current position. Continuously, the behavior of $Object_1$ tracks and to collected information on the time and weather. If the behavior of the observed object through the FMR models to take behavior alerts, Camera will be closely monitored to observe the behavior of $Object_1$. Finally, the behavior of the object by using an association rule algorithm and classification algorithm classifies the pattern of the behavior and

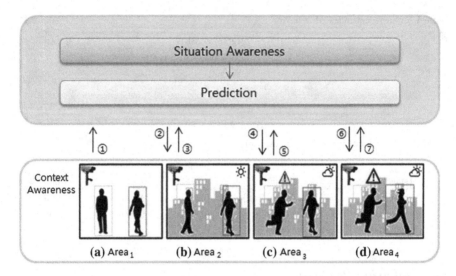

Fig. 3 Implementation result of the Apriori

prediction. When the higher the probability the abnormal behavior gives a notice to another object.

The FMR model analyze the current behavior, by predicting in advance occurs behavioral future, before the crime occurs, it is possible to respond quickly enables proactive.

4   Conclusion

In this paper, CCTV monitoring, as well as the application and use of existing post-processing, a pre-defined abnormal behavior of pedestrians and proposed classification and predictive analytic models to influence the final decision and behavior through the analysis of data. Using Flexible Multi-level Regression models were established assessment rule for behavior. This was detected by observing and predicting the behavior more than the behavior of objects from the CCTV. The proposed model is to reduce the cost and loss of resources. It can also predict the behavior of an object before a crime occurs to quickly determine the situation and decision-making. And it is thought to be a suitable model for crime prevention in advance. However, there remains the following further study. It will establish a continuous model for behavior analysis to discern the abnormal behavior by utilizing environmental factors. Classification algorithm will study them to collected and analyzed about object behavior.

Acknowledgements This work was supported by Institute for Information and communications Technology Promotion (IITP) grant funded by the Korea government (MSIP) (B0101-15-1282-00010002, Suspicious pedestrian tracking using multiple fixed cameras).

References

1. Bark HM (2012) Concept and features of random crime: crimes against random people. J Korean Assoc Public Saf Crim Just 50:226–258
2. Chang IS, Cha HH, Park GM, Lee KJ, Kim SK, Cha JS (2009) A study of scenario and trends in intelligent surveillance camera. J Intell Transp Syst 24:93–101
3. Cho KH, Park HC (2011) A study insignificant rules discovery in association rule mining. J Korean Data Inf Sci Soc 22:81–88
4. Edward TH (1963) A system for the notation of proxemic behaviour. J Am Anthropol 65:1003–1026
5. Hahsler M, Grun B, Hornik K, Buchta C (2009) Introduction to a rules—a computational environment for mining association rules and frequent item sets. The Comprehensive R Archive Network, USA
6. Hong SY (2000) Criminal psychology. Hakjisa, Korea
7. Kim CK, Kang IJ, Park DH, Kim SS (2014) Analysis of the five major crime utilizing the correlation regression analysis with GIS. J Korean Soc Geosp Inf Syst 22:71–77
8. Park SH, Jung SH (2013) A preliminary research on technology development to ensure the safety of pedestrian. National Disaster Management Institute, Korea

Automatic Lighting Control Middleware System Controlled by User's Emotion Based on EEG

SoYoung Ahn, DongKyoo Shin, DongIl Shin and ChulGyun Park

Abstract Recently, human-centered technology development is in the limelight. It reflects user characteristics, including the physical and psychological characteristics of humans. In particular, numerous intelligent systems have been developed to prove humans with the services required for a living space. In this study, a middleware system was developed to analyze a user's emotions using brain waves to control the brightness and color of light accordingly. The middleware in a smart building analyzes the brain waves acquired from the sensors of each household and uses this information to appropriately control the brightness and color of the light in the proper space. Such a system could be utilized in a variety of fields. For example, an intelligent apartment could provide a comfortable indoor environment, as well as save energy, and light therapy could be used to treat depression and insomniac.

Keywords EEG · Illumination · LED lighting · User emotion

1 Introduction

Recently, human-centered technology development is in the limelight. It reflects user characteristics, including the physical and psychological characteristics of humans. In particular, numerous intelligent systems have been developed to prove

S. Ahn · D. Shin · D. Shin (✉) · C. Park
Sejong University, Yulgokgwan 505, 209 Neungdong-Ro,
Gwangjin-Gu, Seoul, Korea
e-mail: dshin@sejong.ac.kr

S. Ahn
e-mail: asy131123@gmail.com

D. Shin
e-mail: shindk@sejong.ac.kr

C. Park
e-mail: cgpark@ecomaytek.co.kr

© Springer Science+Business Media Singapore 2016
J.J.(Jong Hyuk) Park et al. (eds.), *Advances in Parallel and Distributed Computing
and Ubiquitous Services*, Lecture Notes in Electrical Engineering 368,
DOI 10.1007/978-981-10-0068-3_18

humans with the services required for a living space. These intelligent services can save energy in response to a user's request in real time, and create a convenient and fun future living environment [1].

The LED smart lighting market and several related fields currently being developed can maximize the energy saving effect of LED lighting. Moreover, it is possible to provide a reaction that affects human emotions [2, 3]. Reflecting this trend, foreign advanced enterprises are already developing an eco-friendly LED, and our country has also responded quickly to market changes [4].

This study was based on the existing research results showing that the color and brightness of an LED light influence a human's emotions and status. We propose an emotional lighting control middleware system for smart buildings. The system acquires the brain waves of biological signals and determines the user's emotion using the concentration, depression, and relaxation indexes. It can then control the brightness and color of the light automatically according to the user's emotion.

2 Related Research

Studies using the brightness and color of light to control the feelings of people are already being conducted. Shin et al. investigated whether the illumination stimuli of LED lighting could enhance the attention and relaxation level by controlling the color temperature and illuminance level according to activities [5]. In addition, Hoffmann et al. studied the different subjective moods felt by a user under different lighting conditions in an experimental office space [6]. Han and Kim implemented an LED sensitivity illumination system that was driven in response to changes in biological signals, namely the GSR (galvanic skin response) and PPG (photo-plethysmography) signals. After measuring the biological signals from a human body using GSR and PPG sensor modules, the system helped the subject reach a normal state by changing the LED illumination color, corresponding to the state of the subject [7]. In addition, numerous studies on the effects of lighting on emotion have been conducted using the emotions of a human [8, 9].

In this paper, we propose an emotional lighting control middleware system for real-world smart-buildings application on the basis of various previous studies. The system analyzes the user's emotion by measuring their brain waves in real time and controls the brightness and color of the light accordingly. It was found that a user's emotion could be determined more accurately using a biological signal, and a system was developed that could be applied in real life such as to smart buildings.

Table 1 lists the appropriate indoor brightness and color values of the light for comfort according to the task.

A control system was designed to induce a stable emotional state by substituting the appropriate light for each task.

Table 1 Appropriate brightness and color of light

Task (state)	Brightness (lux)	Color
Study (concentration)	500–700	Yellow
Rest (relaxation)	70–150	Blue
Normal (normal)	150–300	Green
Depression (depression)	10,000	Red

3 Emotional Lighting Control Middleware System

3.1 Structure of System

After analyzing the user's emotion by measuring their brain waves, the system developed in this study is configured to control the brightness and color of the light accordingly. First, it collects the normal brain waves of the user and calculates the averages of the indexes. The measured brain waves are analyzed in real time based on the average of each index. The brightness and color of the light are optimized for the four emotions: concentration, relaxation, normal, and depression. Figure 1 shows the structure of the system.

The measured brain waves are separated into multiple waves using a fast Fourier transform (FFT) analysis. Then, the emotion is determined by comparing the averages of the indexes profiled to calculate each index. It thus determines the user's emotion and controls the brightness and color of the light.

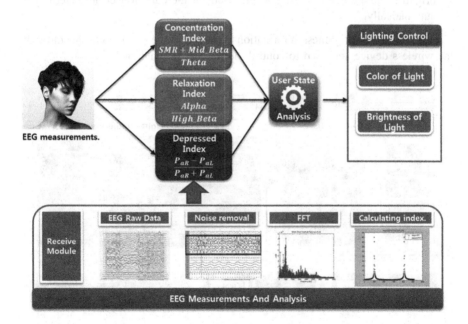

Fig. 1 Structure of system

3.2   Emotional Lighting Control Middleware System Application

The proposed middleware system was developed to provide user personalization and emotional lighting control in the context of households with various members. The input data system of the middleware has a sensor device to obtain the user's emotional information and environment information. First, the emotional information is obtained from EEG devices and sensor equipment based on bio-signals, while the environment information uses environmental sensors such as those for the temperature, humidity, and light intensity (Fig. 2).

The middleware calculates the user EQ (Emotional Quotient) to synthesize bio-information and environmental information and finally uses this to implement light for the emotion corresponding color. The data flow of the middleware for emotional lighting control is as follows.

1. Biological and environmental stream data acquired in real-time are purified through the Metadata Extractor.
2. The purified data are stored in the User Metadata Registry and Emotion Data Warehouse.
3. The environmental situation and emotional state of the user based on the personal ID separated by bio-signals and wearable devices are organized by Multi-level Reasoning.
4. Building a personalized profile on the basis of the extracted information, the brightness and color of the light optimized for the current user are determined automatically.

Optimum color/brightness information is passed to the LED controller through the wireless device and used to control the indoor and outdoor light.

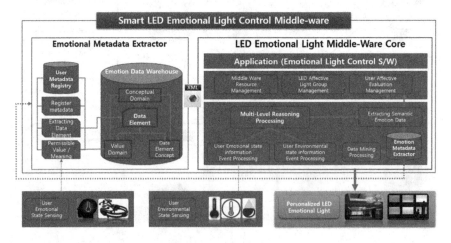

Fig. 2 Middleware framework for emotional lighting control

4 Experiment and Results

4.1 EEG Analysis

An EEG waveform can be represented by a number of separate sine waves or cosine waves with different amplitudes and cycles. Based on the Fourier theory, we used a Fast Fourier Transform analysis method to separate multiple waves. The analysis processing method used the following formula (1).

$$H(f_n) = \sum_{k=0}^{N-1} h_k e^{-\frac{j2\pi kn}{N}} = H_n$$

$$h_k = \frac{1}{N} \sum_{n=0}^{N-1} H_n e^{-2\pi kn/N} \tag{1}$$

We converted the time series signal change over time to a frequency domain signal and determined the pattern of the signal with respect to frequency changes. Using this analysis method, the data were classified using the frequency composition, and the density and distribution of the frequency components were classified. The concentration, depression, and relaxation indexes were calculated using the previously generated information. These formulas are shown in (2)–(4).

$$\text{Power Ratio of } (SMR + Mid\_Beta)/\theta. \tag{2}$$

$$Alpha/High\_Beta. \tag{3}$$

$$(R\_Alpha - L\_Alpha)/(R\_Alpha + L\_Alpha). \tag{4}$$

The user's emotion is classified by setting the tolerance range of the calculated indexes. If the result does not belong to any of the other three (concentration, relaxation, and depression), it is considered to be a normal state.

4.2 Experiment

An experiment was conducted over 5 days on a male in his 20 s. The EEG measuring device used in the experiment was Emotiv's EPOC. It is composed of 14 channels and two ground sensors, and receives data at the rate of 128 Hz/s. This experiment obtained the brain waves of each emotion through meditation or problem solving. We applied the developed system on the fifth day based on the EEG measured over the previous 4 days as a baseline. Table 1 lists the results of the experiment.

Table 2 Results of experiment

	Concentration	Relaxation	Depression
Profiled EEG	0.3975	11.6105	0.005563
Measured EEG	0.4388	12.9456	0.008942

As can be seen in Table 2, each state is divided into a similar value. The classification accuracy can be increased by setting the tolerance range for each state.

5 Conclusion

In this study, a middleware system was developed to analyze a user's emotion using brain waves and control the brightness and color of a light accordingly. The middleware in a smart building analyzes the brain waves acquired from the sensors in each household and controls the brightness and color of the light appropriately. Such a system could be utilized in a variety of fields. For example, an intelligent apartment could provide a comfortable indoor environment, as well as save energy, and light therapy could be used to treat depression and insomniac. In the u-healthcare field, this could be used for the convergence of various medical devices. For example, there are method for treating skin disease using LEDs and LED solariums for medical use, which require tailored treatment through personal biometric information.

However, classification accuracy for the mental state based on the measured EEG was low. To increase the reliability of such an EEG analysis and more accurately classify the status of the user, we should apply a nonlinear algorithm such as SVM or HMM based on the probability.

Acknowledgments This research was supported by Basic Science Research Program through the National Research Foundation of Korea (NRF) funded by the Ministry of Education (2013R1A1A2011350).

References

1. Kim B-C, Choi J-S (2008) A study on implementing a emotional space through virtual models and sounds. Ergonomics Society of Korea, pp 32–33
2. Kwon S-M, Jong I-B (2012) Sensitivity lighting system based on multimodal. J Korea Inst Inf Commun Eng 721–729
3. Kim K-S, Choi A-S (2012) A preliminary study using literature review on the lighting design considering the circadian rhythm. Korean Soc Living Environ Syst 163–170
4. Kim C-H (2009) LED lighting, to be the future of light. In: LG business insight, pp 2–19
5. Shin J-Y, Chun S-Y, Lee C-S (2013) Analysis of the effect on attention and relaxation level by correlated color temperature and illuminance of LED lighting using EEG signal. Technical

report, Journal of the Korean Institute of Illuminating and Electrical Installation Engineers, pp 9–17

6. Hoffmanna G, Guflera V, Griesmacherb A, Bartenbachc C, Canazeic M, Stagglc S, Schobersbergera W (2008) Effects of variable lighting intensities and colour temperatures on sulphatoxymelatonin and subjective mood in an experimental office workplace. Elsevier, pp 719–728

7. Han Y-O, Kim D-W (2014) Sensitivity illumination system using biological signal. JKIECS 499–507

8. Lee E-S (2012) The effect of hue of lighting on affective and cognitive response focused on relaxation and attention. Master's thesis

9. Bauhinia CK, McBride-Chang C (2007) Emotion perception for faces and music. Korean J Thinking Problem Solving 57–65

Hand Recognition Method with Kinect

DoYeob Lee, Dongkyoo Shin and Dongil Shin

Abstract Human interaction is related to the development and maintenance of communication. Communication is largely divided into verbal communication and non-verbal communication. Verbal communication involves the use of a word or words. Non-verbal communication is the use of body language. Gestures belong to non-verbal communication. It is possible to represent various types of motion. For this reason, gestures are spotlighted as a means of implementing an NUI/NUX in the field of HCI and HRI. In this paper, using Kinect and the geometric characteristics of the hand, we propose method for recognizing the number of fingers and detecting the hand area. Because Kinect provides a color image and depth image at the same time, it is easy to understand a gesture. The finger number is identified by calculating the length of the outline and central point of the hand.

Keywords Kinect · Hand gesture · Hand region detection · Depth image

1 Introduction

Human relations depend on communication and are developed and maintained through communication. Communication is achieved using verbal and non-verbal method. Verbal communication involves the use of words or writing. Non-verbal communication conveys ideas using body movements such as winking, shaking hands and laughing. Gestures, which are commonly used in daily life, are also a

D. Lee (✉) · D. Shin · D. Shin
Sejong University, Gunja-Dong, Gwangjin-gu, Seoul, Korea
e-mail: dy03615@gmail.com

D. Shin
e-mail: shindk@sejong.ac.kr

D. Shin
e-mail: dshin@sejong.ac.kr

© Springer Science+Business Media Singapore 2016
J.J.(Jong Hyuk) Park et al. (eds.), *Advances in Parallel and Distributed Computing and Ubiquitous Services*, Lecture Notes in Electrical Engineering 368,
DOI 10.1007/978-981-10-0068-3_19

form of non-verbal communication. Gestures are signs that have been used effectively for communicating with words.

Today, with the diffusion of various image display devices studies on new interfaces have been increasing, as replacements for the existing hardware interfaces. One new interface is an intuitive control method that is easy to use and easy to learn. It is a technology that allows a user to interact with a computer using parts of their body. This is called a NUI (Natural User Interface) or NUX (Natural User eXperience). Research on NUI/NUX has been actively pursued in the field of HCI (Human-Computer Interaction) and HRI (Human-Robot Interaction) [1]. Sound and body movements are the tools this interaction. In particular, gestures have been considered in many studies because they are simple and intuitive [2].

Studies on recognizing gestures have been performed using a variety of equipment. A data glove and video camera are prime examples. A data glove makes it possible to accurately measure the shape and motion of a hand [3]. However, it limits the movement of the hand, requires additional correction and is expensive. A recognition considered here. However, separating the background and hands has not been easy with this method. If there is a change in the light, rapid movement of an object or an object with the color of skin, it is difficult to recognize the user's gesture.

However, with the release of Kinect by Microsoft, it has been possible to solve these problems. Kinect provides a color image and depth image at the same time. The depth image is not affected by a decrease in the brightness of the light. In addition, it is resistant to interference by the background environment, and it can be used even if you have a different skin color than the hand. This is because the depth image provides a depth value. This depth value is the distance between Kinect and an object calculated using the infrared camera. The depth value is a small when the distance between Kinect and the object is, a small. A longer distance between Kinect and the object will produce a larger depth value. At this time, the distance between Kinect and the object is the most efficient from 1.2 to 3.5 M.

In this paper, focusing on this point, using Kinect and the geometric characteristics of the hand, we propose a method for recognizing the number of fingers and detecting the hand area. The proposed method detects the hand region using the depth values, and recognizes the number of fingers using the geometrical features of the hand.

2   Related Work

A gesture is an intuitive and simple piece of information that can effectively be used to control a variety of displays [4, 5]. Therefore, studies have been carried out on various method for gesture recognition. Among the studies using an input device such as a camera and Kinect, a method was presented for recognizing the hand region using the skin color, depth value, or skin color and depth value at the same time.

Han recognized the hand area using the skin color [6]. The first step was to recognize the hand region using the skin color in an image input through the camera. However, if the detected hand region included an arm portion, it could not be accurately recognized. The arm parts may be separated using the geometrical features of the hand. When the hand area is found, excluding the arm, focus is given to the center of the hand area. The center of the hand area has the largest pixel value using the distance converted from the hand area. After recognizing the center of the hand, a circle is drawn in this area to identify the number of fingers. The radius of the circle is 1.5 times the distance from the center of the hand to the hand region. After drawing a circle, the overlapping portion of the circle and hand region is explores in a clockwise direction. The angle of the overlapping portion of the two uninterrupted points is calculated, and a value below 10° increases the number of fingers and 25° increases the number of wrist. After the search ends, the finger number of fingers is recognized by counting.

Jagdish detected the hand region using the depth image of Kinect and determined the number of fingers using the OpenNI module [7]. He separated the background and hand using the depth value of the depth image. Typically, when you make a gesture, the distance between Kinect and the hand is the nearest situation. The Kinect depth image provides a depth value, which differs according to the distance between Kinect and an object. Therefore, using the depth image can obtain the hand region. After acquiring the hand region, the center hand is obtained using a distance transform. Then, the palm area is removed, because it is recognized that drawing a circle in the center of the hand makes it possible to find the finger area. In the finger area, the finger number can be recognized by exploring the depth value.

Tao Hongyoung recognized the hand area and identified the number of fingers using the skin color and depth image of Kinect [8]. Two methods were used at the same time because the hand region could be accurately detected using only the skin color. First, the hand region is isolated by setting the threshold to the depth value. Then, the YCrCb detects a region similar to the skin color in the image. A common portion of the regions obtained through the two methods will be recognized as a hand region. The number of fingers identified as the smallest part of the depth value in the finally detected area.

In addition, in order to recognize a gesture, Choi detected the hand based on its geometric characteristics [9]. Cao determined the pose or shape of a hand based on a database of feature vectors, including extracted features [10].

3   Hand Region Detection Algorithm

Figure 1 shows the flowchart of a hand region detection algorithm proposed in this paper. Kinect provides a color image and depth image at the same time. The depth image is obtained through the infrared sensor. The depth image provides a depth value and is used to recognize the gesture in three-dimensional form. In order to recognize a gesture correctly, the background and hands should preferentially be

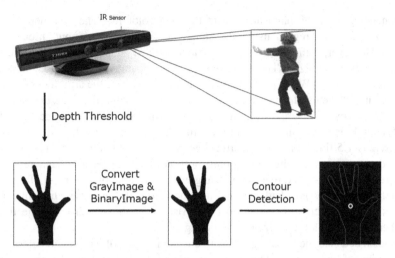

Fig. 1 Proposed algorithm flowchart

removed. These are separated by applying a threshold to the depth values. The depth value is the distance between Kinect and an object and is smaller for a closer distance. After removing the background and hands, in order to recognize the finger number, the depth image is converted to binary images. Then, the outline of the hand area is detected.

3.1 Hand Segmentation

In this paper, a hand recognition method uses Kinect and the geometric features of the hand. First, for correct hand recognition, pre-processing is performed to separate the background and hand. It uses the depth value to separate the background and hand. The depth value is the distance that and objects is from the Kinect and is smaller for a closer distance. Therefore, it obtains the hand region by setting a threshold. The hand region is detected and converted into a gray image and then a binary image, in order to determine the number of fingers. Figure 2 shows the results of detecting the hand area in the image to set a threshold.

Fig. 2 Hand region detected by the threshold setting

3.2 Outlines Detection

Figure 3 shows the result of detecting the center and outline in the detected hand area. The center and outline of the hand are detected using the contour. The contour is a set of pixels that have the same value and includes the coordinates of neighboring pixels. Therefore, the center of the hand is found to calculate the average value of the detected contour coordinates. The outline of the hand is found by connecting the contour coordinates.

3.3 Finger Count Recognition

The number of fingers is detected to using the distance from the outline of the hand to its center. Since the outline consists of a contour, it uses the coordinates of the pixels. Therefore, it can calculate the distance from the coordinates that make up the outline of the hand to the center of the hand. The search direction of the coordinates is counter-clockwise. It compares the successive distance from the coordinates of the outline to the center point of the hand. In a series of three values, if the value of the second is the largest, it is a finger candidate. At this time, the pixels that are below the midpoint of the hand are excluded from the calculation. Figure 4a shows

Fig. 3 The outline and center point of hand

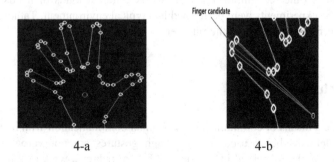

4-a   4-b

Fig. 4 Contour detection and finger navigation process

Table 1 Finger count recognition using the distance

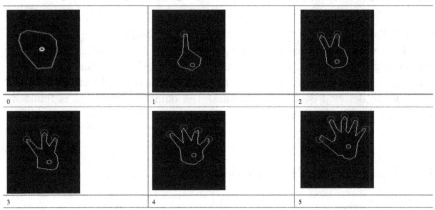

Table 2 Recognition rate according to the number of fingers

Recognition in according to the number of fingers						The average recognition rate
0	1	2	3	4	5	
100 %	100 %	98 %	96 %	98 %	100 %	98.6 %

a contour detection result, and Fig. 4b shows the result of a search for candidates using the proposed method. After the search, you can recognize the number of fingers through the finger candidates. Table 1 lists the results of recognizing the number of fingers.

4 Experiment

The recognition rate was evaluated in order to evaluate the performance of the finger number recognition method presented in the previous chapter. The experiment compared the recognition rates of the six different kinds of hands listed in Table 1. The experiments were performed in a typical environment. Table 2 lists the recognition rates according to the number of fingers.

5 Conclusion

A gesture is a means of communication in human communication with horses, with phonetic elements being used. Even though gestures are non-verbal elements, movements can be expressed in a variety of forms such as a wink or a handshake.

Because of this characteristic, gestures are highlighted as a means of communication between humans and computers. In this paper, we detected the hand region from an image from the input device to recognize a gesture and proposed a method for identifying the number of fingers. The hand area was detected using the depth value provided by Kinect, and finger number was recognized by comparing the coordinates of the center point and the distance of the hand contour that makes up the outline of the hand. In future studies, by increasing the expression range through the recognition of the shape of the hand and finger numbers, the interaction of a human and machine will be easier.

Acknowledgments This research is supported by Seoul R&BD Program (SS11008).

References

1. Wachs JP, Kolsch M, Stern H, Edan Y (2011) Vision-based hand gesture applications. Commun ACM 55:60–71
2. Park SY, Lee EJ (2010) Hand gesture recognition algorithm robust to complex image. J Korea Multimedia Soc 13(7):1000–1015
3. Connelly L, Yicheng J, Toro ML, Stoykov Me, Kenyon RV, Kamper DG (2010) A pneumatic glove and immersive virtual reality environment for hand rehabilitative training after stroke. IEEE Trans Neural Syst Rehabil Eng 18(5):551–559
4. Chen M, Mummert L, Pillai P, Hauptmann A, Sukthankar R (2010) Controlling your TV with gestures. In: Proceedings of the international conferences on multimedia information retrieval, pp 405–408
5. Jain HP, Subramanian A (2010) Real-time upper-body human pose estimation using a depth camera. Technical report, HPL-2010-190, HP Laboratories
6. Han S, Choi J, Park J-I (2013) Two-hand based interaction method using a hybrid camera. In: Proceedings of the of IPIU'13
7. Raheja JL, Chaudhary A, Singal K (2011) Tracking of fingertips and centers of palm using KINECT. In: Proceedings of the 2011 third international conference on computational intelligence, modelling and simulation (CIMSiM), pp 248–252
8. Honyong T, Youling Y (2012) Finger tracking and gesture interaction with Kinect. In: Proceedings of the IEEE 12th international conference on computer and information (CIT), pp 214–218
9. Choi J, Park H, Park J-I (2011) Hand shape recognition using distance transform and shape decomposition. In: Proceedings of the ICIP'11, pp 3666–3669
10. Cao C, Sun Y, Li R, Chen L (2011) hand posture recognition via joint feature sparse representation. Opt Eng 50(12):127210

A Study on the Connectivity Patterns of Individuals Within an Informal Communication Network

Somayeh Koohborfardhaghighi, Dae Bum Lee and Juntae Kim

Abstract Organizational communication structure affects the nature of human interactions and information flow which in its own turn can lead to a competitive advantage in the knowledge economy. However, in addition to that, social relationships between individuals in an organization can also be utilized to produce positive returns. In this article we emphasize the role of individual structural importance within an organizational informal communication structure as a mechanism for knowledge flow and speeding up organizational learning. Our experimental results indicate the fact that structural position of individuals within their informal communication networks can help the network members to have a better access to ongoing information exchange processes in the organization. The results of our analyses also show that through an informal communication network of people in the form of scale-free connectivity pattern organizational learning is faster comparing to small-world connectivity style.

Keywords Centrality measures · Informal communication network topology · Organizational learning

1 Introduction

A learning organization can be understood as a complex network in which individuals interact with each other, aiming at some global purpose. It can be considered as a peer-to-peer system (P2P) in which learning occurs during the

S. Koohborfardhaghighi · D.B. Lee · J. Kim (✉)
Department of Computer Science and Engineering, Dongguk University,
Seoul, South Korea
e-mail: jkim@dongguk.edu

S. Koohborfardhaghighi
e-mail: skhaghighi@yahoo.com

D.B. Lee
e-mail: dblee@dongguk.edu

© Springer Science+Business Media Singapore 2016   161
J.J.(Jong Hyuk) Park et al. (eds.), *Advances in Parallel and Distributed Computing and Ubiquitous Services*, Lecture Notes in Electrical Engineering 368,
DOI 10.1007/978-981-10-0068-3_20

interactions of individuals who are located in it. In a learning organization, collective knowledge of the individuals is needed in order for the organization to reach its overall goals. When an organization becomes a learning organization, knowledge application is necessitated to help organizations to retain correct and valuable knowledge. If the workforces of an organization learn more quickly than the workforces of its competitors we can say that the company has a competitive advantage or edge over its rivals.

Despite the creation of new knowledge and process of learning at an individual level, organizational informal communication structure affects the nature of human interactions and information flow. Therefore, in this paper, we demonstrate how social network of people with different structures may contribute in organizational learning. In this regard, we test organizational learning performance under small-world [1] and scale-free [2] networks in which individuals utilize social influences of special entities in their networks for information exchange processes. We are interested to investigate whether a certain complex network topology would speed up organizational learning or not. We are also interested to check how information diffuses through individual structural importance within a communication network.

The rest of this paper is organized as follows. In Sect. 2, we discuss related works and theoretical background on the topic. In Sect. 3, we detail the model and its parameters. Experimental setup and results are presented in Sects. 4 and 5 respectively. Finally we present our conclusion in Sect. 6.

2 Theoretical Background

In the literature organizational structure is in the list of 11 critical success factors of knowledge management's successful implementation. Both formal communication (organizational structure) and informal communication network structures play an important role in information exchange.

2.1 Organizational Learning and Organizational Communication Structure

Organizations are more successful if they learn sooner and faster than their competitors. That is why the learning concept is growing rapidly and organizations employ it as a competitive advantage. In the organizational learning theory an organization is considered as an adaptive system which is able to sense changes from the environment and evolve to produce the desired outcome. Therefore, a learning organization actively create, store, transfer and use the knowledge for its adaptation to the changing environment. In the literature we can find various proposed models for facilitating organizational learning. In this article we follow the March's model [3] in organizational learning. The detail description of the model is presented in Sect. 3.

The organizational communication structure can be defined as the combination of both formal communication and informal communication network structures. They both play an important role in information exchange. The organizational formal structure is important because it is an indicator of various roles, the hierarchy of these roles and also the distribution of power and authority within an organization. Regardless of the grouping criteria, finally we encounter a meaningful structure through which people communicate with each other. Within such a communication structure we can observe the collective learning which associates internal as well as external learning in the organization. The learning is the combination of Exploration and Exploitation processes. Knowledge exploitation happens through interpersonal learning (P2P interactions) and an organizational communication structure that certainly affect the possible range of solution space.

We followed the same experimental setup in Fang et al. [4] and the details are presented in the experimental setup section.

3 Model

3.1 Entities and Their Individual Structural Importance

Similar to March's model in organizational learning [3], our model has three main entities: an external reality, individuals, and an organization with small-world and scale-free connectivity structures under study. External reality is the organizational goal which is described with a binary vector having m dimensions, each of which has a value of 1 or −1. Values are randomly assigned with the probability of 0.5 for each value in each dimension. There are n individuals in the organization. Each of them holds m beliefs about the corresponding elements of reality at each time step. Each belief for an individual has a value of 1, 0, or −1. A value of 0 means an individual is not sure of whether 1 or −1 represents the reality.

Among commonly used centrality measures eigenvector and betweenness centrality measures are used as a strategic or expert type of knowledge transfer within the network. We select a small-world network, which has a small diameter due to the existence of bridging points. The scale-free network is selected based on the idea that distribution of individual eigenvector centrality followed a power function. Therefore, we are able to observe the existence of hubs and lateral connection in scale-free networks and bridging points in small-world networks. We utilize such centrality information for the diffusion of the reality in second part of our experiments where the goal is to check how fast information can propagated from such positions.

4 Experimental Setup

We generate ten suits of experiments (T1, T2 … T10) to check the performance of learning organization while individuals exchange the information, and we report the average result over 100 simulations run. There are n individuals in each organization (i.e., n = 250). The number of dimensions in the beliefs is set to 100 (i.e., m = 100). Reality is determined by randomly assigning a value of 1 or −1 for each of 100 dimensions, whereas each dimension of an individual's belief set is determined by assigning a value randomly drawn from 1, 0, or −1. Each organization consists of c clusters (i.e., c = 1 in this study) of individuals. The average degree of both networks is set to 4. Similar to Fang et al. [4] model the learning probability is set to 0.3. We adopt generalized learning model of March [3] developed by Fang et al. [4] which is presented below:

$$\phi(x) = s\left(\prod_{j=1}^{s}\delta_j + \prod_{j=s+1}^{2s}\delta_j + \cdots + \prod_{j=m-s+1}^{m}\delta_j\right)$$

Let x_j denote j_{th} element of the bit string x. Then, the linear payoff function can be calculated according to the above formula where $\delta_{j=1}$ if x_j corresponds with reality on that dimension; $\delta_{j=0}$ otherwise. In our formulation we set its value to 5.

5 Results

We compared the organization learning performance of scale-free and small-world networks over 10 suits of experiments and the average result based upon 100 runs is shown in Fig. 1. For each run of the model, we computed the average payoffs of the population during each period t = 0…T. Equilibrium occurred when all the individuals have equivalent knowledge levels. Our first observation in Fig. 1a, b was that the organizational learning performance with small-world connectivity pattern is slower than the scale-free one. Despite the fact that at the end of the simulation the learning curves are very close to each other, the early stage of the simulations is a good indicator of how different network topologies respond to March's payoff function. This clarifies our statement about the fact that in addition to having a realistic model of interactions, with a proper communication network structure organizational learning can be speed up. We were also interested to check how information diffuses through individual structural importance within a network. Therefore, we diffused the reality through individuals with high eigenvector centrality and betweenness centrality values in both networks. The result of our analysis is shown in Fig. 1b. As the result shows, the reality diffuses faster in scale-free network than small-world one. Although at the end of the simulation all

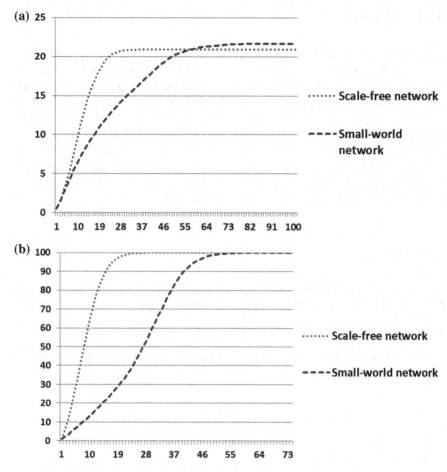

Fig. 1 Informal communication network structure and its impact on organizational learning. a Informal communication network structure: Barabási-Albert Scale-free network and Watts-Strogatz Small-world network, C = 1, n = 250, Plearning = 0.3. **b** Informal communication network structure: Barabási-Albert Scale-free network and Watts-Strogatz Small-world network, C = 1, n = 250, Plearning = 0.3. Reality is diffused by experts (nodes with high betweenness centrality and high eigenvector centrality in Small-world and Scale-free networks respectively) within the organization

the individuals managed to update dimensions of their beliefs, but it seems that in scale-free network the diversity of belief sets is lower in the early stage of the simulation. That means, the organization with a scale-free topology within its communication structure has achieved a high level of eventual knowledge in a shorter time than the one with a small-world topology. Therefore, we can conclude that certain connectivity patterns among the workforces can speedup organizational learning.

6   Conclusion

Distribution and transfer of knowledge is an important part in the knowledge management process. Obtained knowledge within the organization should be generalized and be available to the others through social interactions. We argued that the social relationships between individuals in an organization can be utilized to produce positive returns. The results of our analyses show that organizational learning through an informal communication network of people in the form of scale-free connectivity pattern is faster comparing to the small-world connectivity style. Therefore, a more flexible structure among the individuals within an organization allows the existence of hubs and lateral connections and consequently organizational learning performance can be increased.

References

1. Watts DJ, Strogatz SH (1998) Collective dynamics of 'small-world' networks. Nature 393:440
2. Barabási A-L, Albert R (1999) Emergence of scaling in random networks. Science 286: 509–512
3. March JG (1991) Exploration and exploitation in organizational learning. Organ Sci 2(1):71–87
4. Fang C, Lee J, Schilling MA (2010) Balancing exploration and exploitation through structural design: the isolation of subgroups and organizational learning. Organ Sci 21(3):625–642

Grid Connected Photovoltaic System Using Inverter

HyunJong Kim, Moon-Taek Cho and Kab-Soo Kim

Abstract In this paper, a boost chopper using PV (Photovoltaics) system and PWM (Pulse Width Modulation) voltage type power converter were constructed to provide a pleasant environment to the patients in the hospital wards by controlling temperature, humidity and air-conditioning and heating. For the stable modulation of solar cell, synchronizing signal and control signal were processed using one chip microprocessor. In this thesis, in addition, grid voltage was detected and this grid voltage and inverter output were operated at the same phase for the phase locking with PWM voltage source inverter so that surplus power could be linked to grid. This characteristic were applied on the temperature and humidity sensors in the general buildings and buildings having specific purposes such as hospitals. The good dynamic characteristic of inverter could be obtained by these applications. Also, PWM voltage inverter maintains a high power factor and low-frequency harmonic output so that power can be supplied in the load as well as system.

Keywords Boost chopper · PV · PWM · Inverter

H. Kim
Yeoju Institute of Technology, 338 Sejong-ro, Yeoju-si,
Gyeonggi-do, South Korea
e-mail: yjkfc@yit.ac.kr

M.-T. Cho (✉)
Department of Electrical & Electronic Engineering, Daewon University College,
316 Daehak Road, Jechen-si, Chungbuk, South Korea
e-mail: mtcho@mail.daewon.ac.kr

K.-S. Kim
Aaia Cement Co., 14 Songhaksan-ro, Jechen-si, Chungbuk, South Korea
e-mail: kim97713@naver.com

© Springer Science+Business Media Singapore 2016   167
J.J.(Jong Hyuk) Park et al. (eds.), *Advances in Parallel and Distributed Computing
and Ubiquitous Services*, Lecture Notes in Electrical Engineering 368,
DOI 10.1007/978-981-10-0068-3_21

1 Introduction

Photovoltaics system is categorized into two types according to linking method with utility line. A parallel connection system refers to a system wherein photon is always electrically connected [1]. Whereas, a grid change-over system refers to a system which enables reverse power transmission of surplus power which is generated by photovoltaics. It is always electrically separated and is connected only when generated output is in shortage [1, 2]. In this system, reverse power transmission is not possible and it supplies power only on the load. With these mechanisms, PV system and electric power system have close relationship. Therefore, when PV system is connected to grid, countermeasures are required for output changes of grid system and grid power quality which is hampered by high frequency wave generation, voltage disturbance, and individual drive.

Since output characteristic of solar cell is greatly changed according to insolation and load, it is required to always track maximum output point regardless of insolation and load.

Also, in case PV system is constructed as a stand-alone type, output voltage of voltage source inverter has to be maintained as constant.

When PV system is applied in the houses and small-scale loads, these systems are largely relied on the area and weather. Therefore, in the present study, it was intended to develop an energy saving type source combined power supply unit to obtain power saving effect by around 10–20 % by linking it with utility line to overcome shortcomings of not generating power continuously and independently.

In this thesis, we intended to control boost chopper so that maximum output point can be always tracked regardless of insolation and temperature changes by changing time ratio based on the power comparison after constructing a grid connected photovoltaics system as voltage type inverter. Also, inverter was controlled as phase driving the grid voltage and inverter output by detecting grid voltage in order to synchronize phase so that power of high power factor and low-frequency harmonic were supplied to the load and system.

Besides, supplied power was not used for maintaining temperature and humidity control in the wards and general chambers in the hospital. Instead, photovoltaics system was used so that grid voltage and inverter output were to be driven at the same phase for the separate PWM voltage type inverter and phase locking.

2 Parameter of PV Cell and I-V Characteristic Curves of Solar Cell

Output voltage of a solar battery has almost uniform light and Open Circuit voltage of a solar battery has output of 0.5–0.6 V. Short Circuit Current increases linearly about light and it is reasoned that charging carrier produced by light is proportional to light [3]. Thus, Short Circuit Current of solar battery is very useful about measuring illuminance.

As increasing surface of a solar battery, light reached area that increases area of n semiconductor in wave field region of p-n junction.

Because of Electromagnetic Coupling removes or sends, it increases active region of solar battery. Thus, the ability of supplying power increases.

Output voltage and output current of a Solar battery depends on changing of load in fixed illuminance. The result is caused V-I characteristic curve.

2.1 Short-Circuit Current

Short-circuit current is approximately 5–15 % that is higher than MPP current. A Standard crystalline structure cell (10 cm × 10 cm) has the value of I_{sc} with around 3A under Standard Test Conditions (STC).

2.2 Maximum Power Point

ShortMPP is written in VMPP, IMPP and PMPP Smart Grid Photovoltaic Generation Trainer Value of voltage and current is Nominal in a solar battery. Nominal current Short-circuit current are little different, and solar battery has Short-circuit current internal force. Solar battery also can be phase of Short-circuit [4]. The Fig. 1 represents MPP characteristic curve.

An Solar battery, caused p-n junction without light, generally operates like a semiconductor diode as passive element. The Solar battery with light is converted to active elements, and diode characteristic curve changes. The shape of V-I characteristic stands equally regardless of light.

When temperature increases, active barrier layer is getting thin, thus, Open circuit voltage of the Solar battery decreases to 2 mV/K, it is about 0.4 %/K. If current is supplied by Solar battery, voltage is determined by photoelectric current.

Fig. 1 MPP characteristic curve

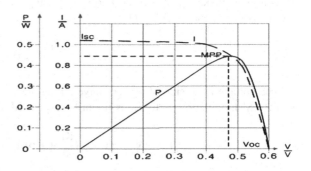

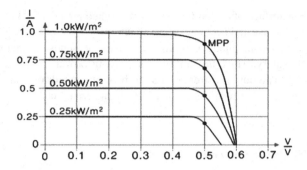

Fig. 2 MPP curves according to light amount

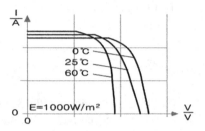

Fig. 3 Temperature characteristic

Therefore, Short circuit current increases about 0.06 %/K and battery power decreases about 0.5 %/K. When temperature increases, efficiency of the Solar battery decreases (Figs. 2 and 3).

3 Control of Grid Connected Power Converter

PI controller makes current of desired size flowing by controlling voltage at both terminals of reactor as in Eq. (1).

$$V_L = V - E = XI^* + (\frac{Ki}{s} + Kp)(I^* - I) \tag{1}$$

The voltage at ac side of power converter can be freely changed by dc voltage and PWM modulation factor, thus it can be expressed as in Eq. (2).

$$E = (V - XI^*) - (\frac{Ki}{s} + Kp)(I^* - I) \tag{2}$$

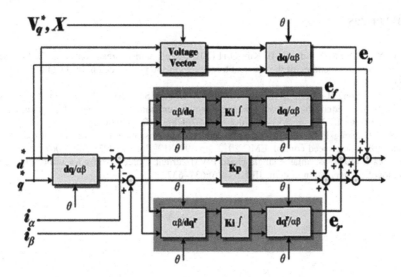

Fig. 4 Block diagram of current control

The first item in right side in Eq. (2) is the voltage after subtracting reactance drop from source voltage, thus it means voltage vector of power converter under steady state. The second item of right side controls the reactance voltage by PI controller so that set current is maintained.

Unbalanced compensation is carried out by current control by the second item of right side in Eq. (2). Proportional control is computed in the stationary reference frame as in Fig. 4. While, integral control is separated into positive phase-sequence component and negative-sequence component and these two rotating reference frames can be expressed as in Eq. (3) as rotating towards opposite direction each other.

$$\begin{bmatrix} e_{vd} \\ e_{vq} \end{bmatrix} = \begin{bmatrix} -Xi_{q*} \\ V_q \end{bmatrix} \tag{3}$$

4 Conclusion

In the present study, PV tracking system was constructed for air-conditioning and heating in the hospital wards. Since output of solar cell which is a dc source is relatively low, PWM voltage source inverter was constructed using a boost chopper having a low capacity dc voltage to operate all the loads inside of hospital including temperature sensor and humidity sensor.

References

1. Kim Y-C, Cho M-T, Song H-B, Kim O-H (2013) Regeneration break control in the hig-speed area using the expending of the constant torque region and power region. Int J Control Autom 6 (4):347–356
2. Kim Y-C, Song H-B, Cho M-T, Lee C-S, Kim O-H, Park S-Y (2012) A study on the improved stability of inverter through history management of semiconductor elements for power supply. GST 2012, CCIS 340, pp 155–162. doi:10.1007/978-3-642-35267-6_20
3. Kim Y-C, Song H-B, Cho M-T, Moon S-H (2012) A study on vector control system for induction motor speed control. EMC 2012, pp 599–812. doi:10.1007/978-94-007-5076-0_73
4. Nonaka S (1994) A suitable single-phase PWM current source inverter for utility interactive photovoltaic generation system. JIEE 114(6):631–637

The Cluster Algorithm for Time-Varying Nonlinear System with a Model Uncertainty

Jong-Suk Lee and Jong-Sup Lee

Abstract In this paper, we discuss the problem of the input-output linearization method for the time-varying nonlinear systems with model uncertainties. The uncertainty of model is presented in all systems, the primary concern in the system analysis. In particular, in this study, while proving the input-output linearization theorem for time-varying nonlinear systems with uncertainty, the uncertainty bounded range of models is derived. It can be seen that the expansion of the input-output linearization for the time-invariant nonlinear system with uncertainty. The Algorithm is simulated and tested by MATLAB. The results show that this algorithm which is more effective routing protocol prolongs the network lifetime.

Keywords Nonlinear time-varying system · Model uncertainty · Input/output linearization · Model matching · Differential geometry · Cloud cluster algorithm

1 Introduction

In most cases the modeling for the natural phenomena, the representation of the system is represented with the time-varying non-linear system. For example, many systems—being operated aircraft, missiles past, rocket flight—are nonlinear models that vary depending on the time. In order to analysis the nonlinear time-varying system, a number of studies have been made. A method for a modeling method of the nonlinear time-varying system is as follows. First, it is to consider the actual

J.-S. Lee (✉)
Department of Business Administration, Joongbu University,
Chubu-myeon, Geumsan-nun, Chungnam, South Korea
e-mail: leejsok@hanmail.net

J.-S. Lee
Department of Culture, Kwangwoon University,
20 Kwangwoon-ro, Nowon-gu, Seoul, South Korea
e-mail: jsleearmy@kw.ac.kr

© Springer Science+Business Media Singapore 2016   173
J.J.(Jong Hyuk) Park et al. (eds.), *Advances in Parallel and Distributed Computing and Ubiquitous Services*, Lecture Notes in Electrical Engineering 368,
DOI 10.1007/978-981-10-0068-3_22

system around the equilibrium point to a linear system. And the method for processing a method for modeling a non-linear time-invariant system with time-varying linear system to screen directly nominal as a function of time. Outside the equilibrium point method and the development of a non-linear system that can interpret Taylor's series to the system in a way easy to ignore the second and later to obtain the linear system obtained in the first equilibrium point has the disadvantage that cannot be taken into account [1, 2].

2   The Input-Output Linearization for the Time-Varying Nonlinear System with an Uncertainty

We let us consider the single input single output time-varying nonlinear system with the model uncertainty.

$$\hat{\dot{x}}(t) = \hat{f}(x(t), t) + \hat{g}(x(t), t)u(t) \tag{1}$$

$$y = h(x(t),\, t) \tag{2}$$

where, the output function $h(x(t),\, t)$ is the smooth scalar field, and assume it not exist the uncertainty in model. The state x is n vector, the state function $\hat{f}$ and input function $\hat{g}$ is n smooth vector field, u and y represented the input and output.

And, it is considered the nominal time-varying nonlinear system corresponding to the Eq. (2).

$$\dot{x}(t) = f(x(t), t) + g(x(t),\, t)u(t) \tag{3}$$

$$y = h(x(t), t) \tag{4}$$

Let a relationship between the time-varying nonlinear system with the uncertainties Eq. (2) and the nominal time-varying nonlinear system Eq. (3) are defined as follow.

$$f(x) = \hat{f}(x(t), t) - f(x(t), t) \tag{5}$$

$$\Delta g(x) = \hat{g}(x(t), t) - g(x(t), t) \tag{6}$$

We can verify the proposed technique by using an example. The nominal time-varying nonlinear system and the model uncertainty is consider as follow. Here α is non-zero real number, output function does not include the uncertainty.

$$\dot{x}_1 = \sin(x_2) + \sqrt{(t+1)}\,x_2$$
$$\dot{x}_2 = x_1^4 \cos(x_2) + x_1 x_2 \sin(x_2) + u$$
$$y = x_1$$
$$\Delta g = (0|\alpha), \quad \Delta f = (x_1|x_2)$$

From the reference [1–3], the nominal time-varying nonlinear system is possible to input-output linearization.

$$T_1 = x_1,$$
$$T_2 = (\dot{T}_1) = \sin(x_2) + \sqrt{(t+1)}x_{(2)} = L_F T_1,$$
$$T_3 = (\dot{T}_2) = (\cos(x_2) + \sqrt{(t+1)})((x_1^4 \cos(x_2) + x_1 x_2 \sin(x_2) + \dot{u})) + 1/2(t+1)^{((-1/2))}x_2$$
$$= (\cos(x_2) + \sqrt{(t+1)})((x_1^4 \cos(x_2) + x_1 x_2 \sin(\dot{x}_2)))$$
$$+ 1/2(t+1)((-1)/2)x_2 + (\cos x_2 + \sqrt{(t+1)})u$$
$$= L_F^2 T_1 + L_G L_F T_1 u$$

Using the transformation T, we get the linearized time invariant linear system. And Eqs. (3), (4) is satisfied. Next the input-output linearization is considered for the time-varying nonlinear system with uncertainties. First, when the output function y continuous differentiating with respect to time t as follows.

$$\dot{y} = (\dot{x}_1) = x_1 + \sin x_2 + \sqrt{(t+1)}x_2 = L_{\hat{F}} h(x, t)$$
$$\ddot{y} = (x_1 + \sin x_2 + \sqrt{(t+1)}x_2) + \sqrt{(t+1)} + \cos x_2)([\![x_1]\!]^4 \cos x_2 + x_1 x_2 \sin x_2 + x_2)$$
$$+ 1/2\sqrt{(t+1)}x_2 + \sqrt{(t+1)} + \cos x_2)u = [\![L_{\hat{F}}]\!]^2 h(x, t) + L_{\hat{G}} L_{\hat{F}} h(x, t)u$$

Since the second derivative of the output y is present input u, the relative degree is 2 [3, 4].

3   The Proposed Algorithm

We propose a cluster algorithm for wireless sensor networks based on Residual Energy and Position, while forming the cluster according to residual energy of sensors and fringe position within clustering [5–8]. It is a key to obtain the optimal electing coefficient which decides the electing time of new cluster head. The algorithm for the proposed work is as follows:

Step1:   Initialize network.
Step2:   The cluster head calculates its residual energy, eh,resi-energy
Step3:   If eh,resi-energy is less than the energy difference, Δenergy, a new election is started. Δenergy is defined as

$$\Delta \, energy = \times MaxEnergy \tag{7}$$

where MaxEnergy is the initial residual energy when this node was elected as the cluster head last time, the electing coefficient in percentage. The optimal value is obtained in our algorithm just as shown in [4]. The election procedure is briefly stated as follows:

Step4: seek for the maximum residual energy, MaxResiEnergy, among the nodes within the cluster.

$$MaxResiEnergy = max \ (ei, resi \text{ - } enery) \tag{8}$$

where ei,resi-enery is the residual energy of node i in the cluster, i the serial number of the node, covering all nodes in the cluster, the node of MaxResiEnergy's ID is set to k.

Step5: whether node k is at the fringe position of the cluster. If yes, go to step4, seek for the maximum residual energy, but not include node k;

Step6: If node k is not at the fringe position of the cluster, check its residual energy, Eresi. If Eresi less than threshold value, go to Step4; If not specify the node of MaxResiEnergy as the new cluster head and substitute the MaxEnergy with MaxResiEnergy.

Repeat Step 2–6 until exceeds the assigned value.

4 Simulation

We use: 1. DCP (data communication procedure) to represent lifetime. DCP means the period of a round of data transfer, including data detection, data transmission and data receipt of a sensor nodes in this cluster; 2r. The r means the times of reversion of cluster head [9, 10]. These are shown in Fig. 1.

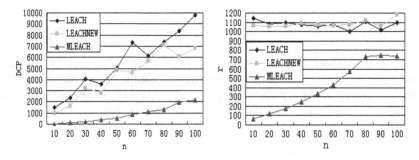

Fig. 1 Comparison of ΔDCP and r last

5 Conclusion

This paper dealt with the input-output linearization problem for the time-varying nonlinear systems with uncertainties. That is, by extending the input-output linearization technique for the time-invariant nonlinear systems with the uncertainty, it is proposed the input-output linearization of the time-varying nonlinear system that includes the uncertainty. Proposed input-output linearization techniques, it contains a conventional time-invariant linearization techniques, also include observability for time-varying nonlinear systems. In this paper, the proposed input-output linearization technique induced a bounded of uncertainties, the proposed scheme is proofed and verified by the example.

References

1. Dabo M, Langlois N, Chafouk H (2009) Dynamic feedback linearization applied to asymptotic tracking, Generalization about the turbocharged diesel engine outputs choice. In: Proceedings of the American control conference ACC'09, pp 3458–3463
2. Lee J-Y, Jung K-D, Hong B, Cho S (2014) Method of extended input/output linearization for the time-varying nonlinear system. Future Inf Tech Lect Notes Electr Eng 309:37–46
3. Lee J-Y, Cho S (2014) Stabilization inverse optimal control of nonlinear systems with structural uncertainty. Adv Comput Sci Appl Lect Notes Electr Eng 279:1343–148
4. Lee J-Y, Jung K, Cho S, Strzelecki M (2014) Nonlinear time-varying control based on differential geometry. Int J Internet Broadcast Commun 6(2):1–9
5. Heinzelman W, Chandrakasan A, Balakrishnan H (2000) Energy-efficient communication protocol for wireless sensor networks. IEEE Computer Society, pp 175–187
6. Yang Y (2006) A wireless sensor network routing protocols based on LEACH. Master thesis, University of Electronic Science and Technology of China, pp 50–68
7. Wang W, Jantsch A (2006) A new algorithm for electing cluster heads based on maximum residual energy, pp 1465–1470
8. Lee JY, Jung KD, Shrestha B, Lee J, Cho S (2014) Energy efficiency improvement of the of a cluster head selection for wireless sensor networks. Int J Smart Home 8(3):9–18
9. Cho S, Shrestha B, Shrestha S, Lee JY, Hong SJ (2014) Energy efficient routing in wireless sensor networks. Int J Adv Smart Converg 3(2):1–5
10. Song YI, Lee JS, Jung KD, Lee JY (2014) Energy efficiency hierarchical multi-hop routing protocol for wireless sensor network. Adv Appl Converg Adv Cult Technol 148

Integrated Plant Growth Measurement System Based on Intelligent Circumstances Recognition

Moon-Taek Cho, Hae-Jong Joo and Euy-Soo Lee

Abstract In this paper, based on core technologies such as overcoming a place's limitations, light that can substitute for the sunlight, automation, nutrient supply system and temperature, and intelligent situation recognition for solar power generation, geothermal HVAC (heating, ventilating, and air conditioning), a plant growth analysis system for vegetation factories was designed. The system is likely to improve the freshness of agricultural products through order and planned productions, to create new markets through the convergence of the IT and BT industries, and to promote convenience in farming and comfort in workspaces through automatic control, robot development, etc. In addition, the system is expected to offer opportunities for urban residents to experience and learn the whole process of a plant's growth; to provide a leisurely life, such as a downtown oasis, to those who are tired of the dreary city life; to prevent environmental pollution through the effective use of recycled resources; and to produce and stably supply diverse agricultural products all year round, regardless of the weather.

Keywords Intelligent situation recognition · Vegetation factory · Plant growth measurement · Integrated plant growth management

M.-T. Cho
Department of Electrical & Electronic Engineering, Daewon University College, 316 Daehak Road, Jechen-si, Chungbuk, South Korea
e-mail: mtcho@mail.daewon.ac.kr

H.-J. Joo (✉) · E.-S. Lee
College of Engineering, Dongguk University, Seoul, South Korea
e-mail: hjjoo@dongguk.edu

E.-S. Lee
e-mail: eslee@dongguk.edu

© Springer Science+Business Media Singapore 2016   179
J.J.(Jong Hyuk) Park et al. (eds.), *Advances in Parallel and Distributed Computing and Ubiquitous Services*, Lecture Notes in Electrical Engineering 368,
DOI 10.1007/978-981-10-0068-3_23

1 Introduction

A plant factory is a plant production system in an artificial ecosystem. It artificially controls the growth environment and produces same-quality produce based on the same production schedules seen in a manufacturing factory. The demand for plant factories is not yet high due to their low profitability resulting from their high construction, operation, and production costs and the non-standardization of the technology. These problems need to be addressed, and this can be done by standardizing the facility, production, and operation [1, 2].

In this study, a growth analysis system of a plant factory was designed based on the ICT analysis of the photovoltaic power generation and ground heat pump system as well as of key elements like place, light, automation, nutrient solution supply system, and temperature. The freshness of the produce was enhanced through production based on orders and plans, new markets were created through the combination of the IT and BT industries, and the convenience of the work and the comfort of the area were improved through automatic control and robot development. The system also provided an opportunity for the city dwellers to observe and study all the courses of the plant growth, and helped them relax considering the stressful city life. It prevented environmental pollution through the effective recycling of resources, and helped construct a system that could stably produce various produce throughout the year, regardless of the climate. While the plant production system in an artificial ecosystem can produce same-quality produce based on production schedules through the artificial control of the growth environment and production automation, it has such problems as high construction and operation costs, cultivar deficiency, non-standardization of the technology, and unstable profitability.

2 Related Studies

2.1 Growth of Plants

Photosynthesis, which secures energy in the form of a chemical from light, is the most important chemical process related to plant growth [3].

A plant secures energy through photosynthesis, as shown in Fig. 1. It grows through respiration and biosynthesis with the various organic matters absorbed by it. Cellular respiration is one of the metabolic processes; it is a complicated chemical reaction in which an organism obtains energy by decomposing organic compounds. These are the factors required for growing plants, and their conditions can determine the quality of the produce. These are some of the elements that must be controlled when forming an artificial ecosystem.

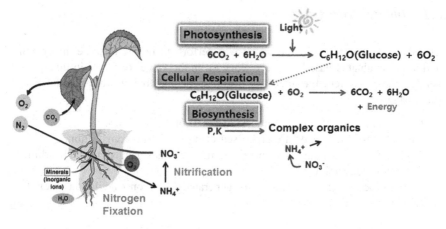

Fig. 1 Metabolism process of a plant

2.2 Growth Measurement

The shapes and physiological functions of the plants are measured in growth measurement. The measurement takes place in real time, and it is important to minimize the stress of the plant during the measurement. Figure 3 shows the growth measurement sensors used.

The sensors positioned in satellites, airplanes, and others are used to measure the status of the crops growing on large areas. The growth of the crops is measured with the sensors by measuring the amount of the reflected light with various wavelengths.

3 Growth Measurement System Based on Intelligent Circumstances Recognition

A database of cultivation circumstances information, video images, and production data was constructed to predict the crop production volume for production planning. Real-time video images, a leaf area analysis program, and middleware were developed. The production volume was predicted and analyzed using the database [4, 5]. The photosynthesis efficiency was determined by measuring the CO_2 consumption with a 200 µmol m^{-2} s^{-1} LED composition from the combination of the red light (630 nm) and the blue light (450 nm), on a one-by-one basis.

Also, the photosynthesis demand and the metabolism of the plants were measured according to the lighting, for the simultaneous cultivation of the plants with similar levels of photosynthesis efficiency. After the measurement of the height, width, and leaf area, variable selection through best subset regression and regression analysis through multivariate analysis were performed.

3.1 Image Processing

The shape measurement of the plants was performed in one to three dimensions, and it was evaluated based on the average or on the value for a certain time period because the growth changes daily [4]. The four methods below were used for image processing.

1. Point processing: Pixels are processed based on their values.
2. Area processing: Videos are processed using mask and convolution.
3. Topological processing: The pixel arrangement is changed through arbitrary geometric conversion.
4. Frame processing: Pixel values are generated based on calculations including two or more different videos.

3.2 Image Analysis

The growth status was diagnosed by capturing images of the selective wavelength range. Figure 2 shows the LUV conversion of the input video and the LUV color analysis, which extracted a wavelength corresponding to the UV value of the pixels.

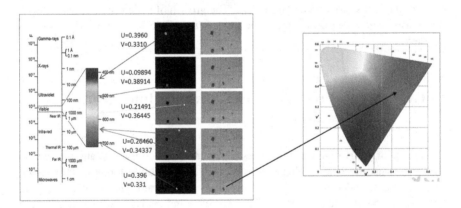

Fig. 2 UV values according to the colors of the color video images and the corresponding wavelengths

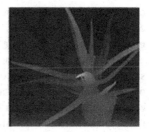

Fig. 3 Disparity map image obtained using a stereo camera

3.3 Three-Dimensional Measurement of a Plant Using a Stereo Camera

Plant shape analysis at each step of growth was used for the germination and seedling. An area algorithm was developed using the proportion of disparity values in the segmentation area. The depth value was measured using the distance between the two matching pixels from the left and right stereo images. Figure 3 is the disparity map image obtained using a stereo camera.

3.4 Machine Learning

Machine analysis is used for segmentation or circumstances recognition in image processing. It has four classifications, as shown below [6, 7].

1. Supervised learning: Analogizes a function from the training data. Generally includes the attribute of the input object in the form of a vector. Displays the wanted results for each vector
2. Unsupervised learning: Analogizes a pattern from data without labels, as in the density estimation of statistics
3. SVM: Judges based on the distance between the support vectors on a hyperplane
4. K-means: Obtains the Hausdorff distance between a normal leaf and a stressed leaf in a reflectivity data group.

4 Conclusion

The existing operational systems in plant factories perform controlling or simple monitoring through the devices installed in the factories. Also, they use devices constructed based on the experience of the developer. Especially, the automated systems for controlling the growth environment in the existing system were constructed in a static structure, and adopted algorithms that did not consider the characteristics of the crops. The entire program thus needs to be revised to apply the growth algorithms of other crops.

While the plant production system in an artificial ecosystem can produce same-quality produce based on production schedules through the artificial control of the growth environment and production automation, it has such problems as high construction and operation costs, cultivar deficiency, non-standardization of technology, and unstable profitability. If these problems are addressed, the system is expected to help boost the farming competitiveness and profits.

References

1. Chung S, Kim HM, Kim SD (2007) Formulation of stable Bacillus subtilis AH18 against temperature fluctuation with highly heat-resistant endospores and micropore inorganic carriers. Appl Microbiol Biotechnol 76:217–224
2. Idris EE, Iglesias DJ, Talon M, Borriss R (2007) Tryptophan-dependent production of indole-3-acetic acid(IAA) affects level of plant growth promotion, Bacillus amyloliquefaciens FZB42. Mol Plant-Microbe Interact 20(619–626):2001
3. Ellis G, Krah JO (2001) Observer-based resolver conversion in industrial servo systems. In: Proceedings of the PCIM 2001 conference, 21 June 2001
4. Hoseinnezhad R, Harding P (2005) A novel hybrid angle tracking observer for resolver to digital conversion. In: Proceedings of the 44th IEEE conference on decision and control, and the European control conference, pp 7020–7025
5. Analog Devices (2003) 12-bit R/D converter with reference oscillator. Analog Devices
6. Devices Analog (1990) Using the ADSP-2100 family volume 1. Analog Dev Rev 1:51–66
7. Texas Instruments (2000) TMS320F240 DSP solution for obtaining resolver angular position and speed. Application report SPRA605

A Study on the Big Data Business Model for the Entrepreneurial Ecosystem of the Creative Economy

Hyesun Kim, Mangyu Choi, Byunghoon Jeon and Hyoungro Kim

Abstract The entrepreneurship required for a creative economy is one that promotes job creation through the creativity of CEOs who have entrepreneurial spirit as well as excellent knowledge and technology. This entrepreneurship plays a critical role in the construction of a venture business ecosystem. Meanwhile, big data, until now, have consisted only of numbers and texts that are specified and standardized by certain structured rules. Nowadays, however, big data analysis methods are being developed to gain information and business opportunities from new aspects through the use of nonstandard data. The big data technology is becoming a core element in the new digital age rather than just a trend, and the big data strategies are progressing from the testing stage to the implementation stage. In particular, as the importance of nonstandard data is increasing, limitations of the conventional system analysis appeared, and the analysis methods of advanced analytics are being highlighted. The scope of big data is anticipated to expand in enterprises with the emergence of many application cases, such as the real-time use of the data. The business model for the formation of the entrepreneurial ecosystem of the creative economy can assist the construction of an entrepreneurial platform as a catalyst for the stimulation of innovative business start-ups. This entrepreneurial platform can produce such effects as fast product development, commercialization of technology, risk reduction, and job creation. In this study, the operating model formation according to the characteristics of the industry or the business model in

H. Kim · M. Choi · B. Jeon
Department of Engineering, Dongguk University, Seoul, South Korea
e-mail: daisyhsun@dongguk.edu

M. Choi
e-mail: mgchoi@dongguk.edu

B. Jeon
e-mail: bhjeon@dongguk.edu

H. Kim (✉)
Kaywon College of Arts, Uiwang-si, Gyeonggi-do, South Korea
e-mail: Hyoungro@hanamil.net

© Springer Science+Business Media Singapore 2016

185

J.J.(Jong Hyuk) Park et al. (eds.), *Advances in Parallel and Distributed Computing and Ubiquitous Services*, Lecture Notes in Electrical Engineering 368,
DOI 10.1007/978-981-10-0068-3_24

operation, the profit creation sources, the scope of value provision, and the priorities with regard to the required data were examined. Furthermore, cases related to the performance of big data business models of advanced corporations were examined, and a big data business model for the formation of the entrepreneurial ecosystem of the creative economy in the 21st century was derived.

Keywords Entrepreneurial ecosystem · Venture · Big data · Statistical analysis

1 Introduction

The interest in the entrepreneurial ecosystem as a driving force behind a creative economy is rising, as is the interest in big data business to create analysis value through the creation and processing of big data. However, the strategies for the entrepreneurial platform are inadequate, in contrast to the efforts made to stimulate the entrepreneurial ecosystem from the big data business perspective. This big data phenomenon is quickly spreading together with the explosive increase of customer data collection activities and multimedia contents, the activation of SNS through the propagation of smartphones, and the expansion of the base of the network of things. Meanwhile, the creative government based on creativity emphasizes Government 3.0, which is defined as "a governance administration system or a government management system that aims to play the role of a cooperative partner by redesigning administration methods and procedures based on highly intelligent ICT and social networks, sharing knowledge and information among the government, companies, citizens, and global communities, and providing a public platform that can continuously create productive and democratic added values through the mutual transactions among the members of the society" [1–3].

The business model for the formation of the entrepreneurial ecosystem of the creative economy enables the construction of an entrepreneurial platform as a catalyst for innovative business start-ups. Such entrepreneurial platform can generate effects like fast product development and commercialization of technology, risk reduction, and job creation. Therefore, in this study, the organization of an operation model, the profit creation sources, the scope of value provision, and the priorities with regard to the required data were examined according to the characteristics of the industry and the type of business model in operation. Furthermore, cases of the big data business model performance of advanced corporations were investigated, and a big data business platform model for the formation of an entrepreneurial ecosystem was derived.

2 Related Research

The global leading enterprises from the perspective of big data include IBM, Oracle, EMC, HP, Google, SAP, SAS, and MS. IBM is dedicated to the attainment of big-data-related technologies through the acquisition of data analysis companies such as Essential, which specializes in data storage management and integration, by investing over 14 billion dollars for the last 5 years [3, 4]. Furthermore, National Information Society Agency [4] emphasizes the importance of national strategies to change the direction of information systems from system orientation to data orientation, and the construction of data infrastructure to respond to the paradigm shift, and asserts that such data infrastructure can act as a core engine to create opportunities while responding to the characteristics of the future society, such as uncertainty, risk, smartness, and convergence. NISA predicts that by 2020, the era of zetta bytes will begin, and the RFID and sensor data types and the real-time characteristics of data will be fully developed.

Big-data Model insists that the purpose, sector, opening, and security of big data are more important than the concept of big data [5, 6]. According to a Mckinsey report [1], big data processing abilities will lead to the improvement of the productivity of enterprises and of the national economy, and to the reinforcement of corporate competitiveness. The productivity improvement effects of big data were also estimated by industry, and it was found that computers, electronic products, and information and communication technology were the areas that had the highest big data value and that promoted innovations the most.

3 Big Data Business Strategies in the Creative Economy

The goal of venture companies in the era of the creative economy is to develop customized services aligned with big data through the establishment of big data technology commercialization strategies, and to reinforce technical competitiveness through the establishment of business strategies based on data. In line with the growth of big data markets, global companies such as Google, Face-Book, Amazon, and Apple are continuously leading the creation of new services by building a virtuous circle in which they collect data again through their own services. IDC predicted that the big data market would grow 10 times, from US$5 billion in 2012 to US$50 billion in 2017.

Therefore, enterprises should first analyze thoroughly their internal and external risks and opportunities, and review their abilities for searching, using, and analyzing the big data that they possess in partnership with big-data-related organizations. Furthermore, all organizations should participate in the big data industry, which is emerging as IT convergence solutions.

The top priority for organizations is to develop data analysis techniques that can integrate, search, and utilize the vast data that they possess, and diagnosis indicators for information exchange. Furthermore, they should develop indicators for big data quality measurement following the diagnosis results, and should feedback the results. In other words, they should use the collection systems and analysis bases of specialist companies or develop algorithms for analyzing data in scientific and creative ways.

Recently, the government stressed that big data, touted as the "oil of the 21st century," is the new capital of the creative economy that creates high added values and jobs with creativity and ideas, and that we should reinforce our industrial competitiveness based on big data and cloud sourcing.

For the growth of the big data industry, we must build the infrastructure that is needed by enterprises, and support big-data-based business start-ups and customized commercialization. In order to build the big data infrastructure, we must attain human and material resources, upgrade the resources through education, etc., and also acquire funds to support strategic research and development. Above all, we must discover domestic and international success stories of big data business, and stimulate exchanges among different enterprises. For this purpose, enterprises should construct a system for acquiring innovative ideas through collective intelligence, sharing results, and commercializing ideas.

The analysis platform for the establishment of big data strategies involves data collection and integration, data preprocessing, data storage management, data

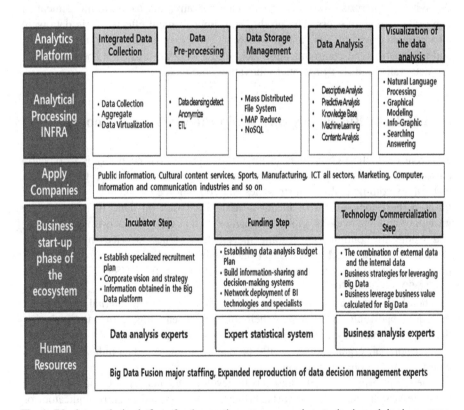

Fig. 1 Big data analysis platform for the creative economy, and strategies in each business stage

analysis, and visualization. Furthermore, the analysis infrastructure involves storage in NoSQL, etc. after data virtualization, and visualization such as graphic modeling through such analysis techniques as contents analysis and descriptive analysis. This is applicable to all industrial fields (Fig. 1).

Data analysis involves the real-time analysis of the data produced by various entities in various environments, and the utilization of data analysis can be enhanced through the derivation and visualization of various patterns through big data processing in data analysis, such as statistics.

Furthermore, to propose big data business steps in line with the entrepreneurial ecosystem, in the incubator stage, we should focus on hiring specialists who collect, handle, and process data, and on acquiring big data platform information. In addition, we should educate data analysis specialists. In the funding stage for the foundation of the big data business, we should analyze the funds required for data analysis and establish funding plans, and we should build an information sharing and decision making system for data analysis.

4 Conclusion

In this study, the organization of an operating model, the profit creation sources, the scope of value provision, and the priorities with regard to the required data were examined according to the type of business model. Furthermore, the big data business model performance of advanced corporations was investigated, and a big data business platform model for the formation of an entrepreneurial ecosystem was derived.

In conclusion, the big data business process in line with the entrepreneurial ecosystem of the creative economy consists of three steps: incubation, funding, and technology commercialization. In each stage, data analysis specialists, statistics system specialists, and big data business analysis specialists need to be educated, and an ongoing education system must be established. Furthermore, it is important to expansively reproduce human resources specializing in convergence areas such as mathematics, statistics, science, IT, and business administration. This is because good data scientists are required to acquire complex and advanced knowledge and abilities in various areas in order to have expertise in many different areas and exercise their abilities in complex ways.

References

1. Joo HJ, Hong BH, Kim SS (2012) Smart-contents visualization of publishing big data using NFC technology. CCIS 351:118–123
2. National IT Industry Promotion Agency (2011) Weekly technical trends, 15 July 2011

3. Korean Agency for Digital Opportunity and Promotion (2011) New value creation engine: the new possibilities of big data and responding strategies
4. National IT Industry Promotion Agency (2012) Weekly technical trends, 11 April 11 2012
5. ITU-T Recommendation I.350 (1993) General aspects of quality of service and network performance in digital networks, including ISDNs, March 1993
6. ITU-T Recommendation Y.1541 (2003) Network performance objectives for IP-based services, February 2003

Implementation of Intelligent Decision-Based Smart Group Scheduler

Kyoung-Sup Kim, Yea-Bok Lee, Yi-Jun Min and Sang-Soo Kim

Abstract Unlike in the past, where diaries and paper calendars were mostly used, the rapid increase in the use of computers and smartphones of late has demanded a function with which the users can confirm and attach their schedules anytime, anywhere. Especially, in the busy and flexible modern life, fast schedule coordination targeting a majority has become increasingly necessary. As the conventional schedule management methods have had problems in coordinating a common schedule that satisfies the majority in cooperation and group work systems, in this study, a smart group scheduling function was developed, with focus on intelligent decision scheduling, unlike in the conventional schedule management programs. Further, the function proposed in this paper, whose process deduces a result shared by majority people in a group, can be concluded within a system based on individuals' opinions and information; therefore, it is considered the system has the potential to contribute to democratic and horizontal work execution.

Keywords SGS (smart group scheduler) · Intelligent decision making · Schedule management

K.-S. Kim · Y.-B. Lee · Y.-J. Min
Department of Computer Science and Engineering, Dongguk University,
Seoul, South Korea
e-mail: kyskks@dongguk.edu

Y.-B. Lee
e-mail: oops6688@hanmail.net

Y.-J. Min
e-mail: jaeheui723@gmail.com

S.-S. Kim (✉)
MLT Co., Ltd, 267 Simindae-Ro, Dongan-gu,
Anyang-si, Gyeonggi-do, South Korea
e-mail: cqsky@paran.com

© Springer Science+Business Media Singapore 2016   191
J.J.(Jong Hyuk) Park et al. (eds.), *Advances in Parallel and Distributed Computing and Ubiquitous Services*, Lecture Notes in Electrical Engineering 368,
DOI 10.1007/978-981-10-0068-3_25

1 Introduction

Not only for company, but communication between individuals has gotten more important as well, cooperation system has been applied for it [1]. For group schedules, we have to put through a series of process such as deciding the date of the meeting and letting group members know as an announcement. Nevertheless, general scheduler software only provides functions for individual, or registration of the schedule. Sometimes overlapping of schedules of group and individual happens. This program increases the effectiveness by lessening required process till registrating schedules and makes reliability high by considering each individual's schedule.

Existing group ware, calender, and SNS provide functions of schedule registration and management. However, it is sometimes inefficient because of its one-sided vote for making group's schedule especially when they decide the date of meeting [2, 3]. Therefore, we use automatic registrating function of schedule and time-searching algorithm. It calculate the space time for all of the group members so that it gives reliability and effectiveness for the decision and expands group ward and social function as well.

Activate the cooperation system and increase the effectiveness of tasks. For getting expandability and the popular appeal, we develop the integrated scheduling program which is viable for the decision making based on the html5. In a bid of the real-time enterprise work process, the final goal for the project is development of scheduling program available on multi devices with html5. For the graduation project in 2015, prior to the final program development, we focus on the main function of the smart group scheduler, realization of the smart scheduling algorithm through searching time space of the group members, and prototype of the final form of program.

2 Related Work

There are various forms of scheduling program existing in the world and to linkage them, IETF provides the calendar standard. It regulates application programs to linkage each other.

RFC 2445 Internet Calendaring and Scheduling Core Object Specification [3].

Many programs provide various functions with linkaging other programs apart from their main scheduling function (Figs. 1 and 2).

HTML5 is the newest form of HTML programming language to make web document. With HTML5, we can realize functions without installing Active X. Especially, it its available to make fancy graphic effects without flash, Silverlight, JAVA FX on the Web browser [4]. Therefore, the strongest advantage of HTML5 is its compatibility. Most of the web applications right now are being developed with HTML5 and CSS techniques.

Fig. 1 Google calendar

Fig. 2 Apple iCal

3 Requirement Analysis

3.1 Function-Related Requirement

Individual schedule registration: Functions for managing each individual's schedule. Group schedule registration: After selecting the group and generating schedules, synchronization is available on each member's calendar in the group. File sharing: File uploading on each schedule. Group members are able to open attached file with schedule. Automatic generation of group schedule: After selecting group, we enter the description for the information of meeting such as title, time of the schedule. It is sent to all group members so that they can also enter each of their available time.

All entered data are calculated and schedule is generated as its result. Location mark: location of meeting is showed when we select the registered schedule. Group management: Add/Delete group.

3.2 Data-Related Requirement

Group schedule should be available to be adjusted by all group members. One simple data which includes all registered users on the program. Individuals can not access to the group data if the group is lifted.

3.3 Interface-Related Requirement

Differentiation between individual and group schedules. Show as monthly, weekly, daily calendar. Size control for the interface of the mobile and the PC. Standard interface for using data of other calendar.

3.4 User-Related Requirement

Compare to existing scheduler program, special operation method should not be needed. Linkage with SNS. Automatic long-in is available when the information is entered initially. Convenience for using data of other calendars.

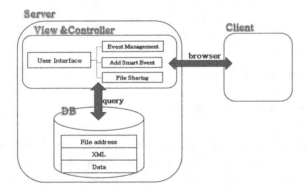

Fig. 3 System structure; *Event Management* Edit, delete the Individual/Group schedule; *Add Smart Event* Check Individual's spare time + Available time → Automatic generation of group's schedule; *File address* Path of file storage when file is shared; *XML* Stores the schedule data as xml

4 System Structure

Smart Scheduling—Computing algorithm is started when all members on group enter their available time in a set time. If not all members on group enter the information in a set time, it still starts computing if more than a half members participated entering, but cancels scheduling with less than a half members (Fig. 3).

5 Conclusion

Among existing program, those of the most distributed programs are Naver calendar, Line calendar. Scheduler application which is supported natively by cell phone device tends to be used widely. Different from the past when we mostly used the diary and the paper calendar, Function to check and add the schedule wherever and whenever has been required as the usage of smart phone is rapidly increased. Since most of modern people are busy and living a flexible life, the importance of instant management for group schedule has been increased day by day. Considering this requirement, we made pro-user scheduler program for individuals and groups. It has been always issued that managing schedule to let multitude be satisfied on cooperation system or team work. Therefore, we specially developed 'Smart Group Scheduling' function distinct from existing scheduler programs. This function is the main part of our program. By moving the decision of schedule from user to the program, it aims to generate more objective and rapid group schedule. We thought this function would be helpful for modern people who are suffering from problem with decision making.

As we looked through many existing scheduler programs, we figured out that those of the most important components for the scheduler program is not its various and fancy functions but pro-user functions such as visibility, simplicity and ease of use. Therefore, we resolutely removed some functions which are still left on existing programs but has low usage, and focused on basic functions of calendar and group scheduling. Since this program requires much inputs from user and interaction, it was hard to assume exception and number of cases. Especially for the group scheduling function, since there are a lot of information to be gathered and algorithm to be accessed as well to draw a result value, we had hard time realizing it. However, we could get over them by repetitive use of same cases, case distribution, enough detail data collected from user. In addition, this program contributed to forming democracy and horizontal-business atmosphere because the process for the result is based on 'all' members' opinion on group rather than 'some' of them.

References

1. Byeong-Gweon G et al (1999) Web-based groupware solution. Korea Inf Process Soc J 6(3):101–109 (1226–9182)
2. Dae-Ryong B et al (2001) Design and implementation of a wired and wireless internet system. In: Korea information processing society, 2001 fall academic conference collection, vol 2. pp 1067–1070
3. Hee-Jong P et al (2003) Events processing for bio place schedule management. In: Korean institute of information scientists and engineers, the 30th KISS Fall Conference in 2003, vol 3, pp 334–336
4. Seong-Soo K (2006) WIPI-based e-mail and schedule management system. J Korea Contents Assoc 6(7):50–57 (1598–4877)

Implementation of MCA Rule Mapper for Cloud Computing Environments

Kyoung-Sup Kim, Joong-il Woo, Jung-Eun Kim and Dong-Soo Park

Abstract The client-server system in the old finance and banking circle was operated in such a way as to respond to a relatively small number of devices, such as ATMs (automated teller machines) and personal computers. Due to the significant increase in the use of smart devices at the present time, however, the number of devices that need to be connected to financial transaction servers continually skyrockets. In this paper, a mapper that can analyze the information of the messages sent from newly added terminals and then link to the MCA was designed, which would allow new terminals to be added without correcting the programs. In addition, as myriads of terminals need to be added, the programs cannot be corrected on a case-to-case basis; in addition, it is actually impossible to correct all the programs. Thus, the mapper, an automatic mapping system, was developed to add terminals automatically. As a result, the mapper is likely to minimize the labor and money that need to be invested for mapping works and to establish a definitive standard in using various mapping tools, thus preventing losses accrued from the use of each different tool.

Keywords MCA · MRM · Delivery channel system · Meta field data · Cloud computing

K.-S. Kim · J. Woo · J.-E. Kim
Department of Computer Science and Engineering,
Dongguk University, Seoul, South Korea
e-mail: kyskks@dongguk.edu

J. Woo
e-mail: lllollipop@naver.com

J.-E. Kim
e-mail: game017@daum.net

D.-S. Park (✉)
NOVOSYS Co., Ltd, 82-1 Pil-dong, Jung-gu, Seoul 100-715, South Korea
e-mail: toward21c@empas.com

© Springer Science+Business Media Singapore 2016   197
J.J.(Jong Hyuk) Park et al. (eds.), *Advances in Parallel and Distributed Computing and Ubiquitous Services*, Lecture Notes in Electrical Engineering 368,
DOI 10.1007/978-981-10-0068-3_26

1 Introduction

In past bank's Client-Server System was operated by responding to relatively few device such as ATMs and personal computers. However devices which want to connect to bank servers increased geometrically as frequency of smart device usage has increased a lot. As a result, it is impossible for few administrators to update MCA server (intermediate server) whenever it needs to be attached new channels. It is the reason why we need automated method to update MCA server; MCA Rule Mapper [1, 2].

Recently, increase of interest of integrated channel system, MCA and MCI are becoming a buzzwords in industry. In early and midterm 90's, banks only fully committed to delivery channel and nexus of clients. But they knew each-constructed delivery channel had a lots of problems [3]. To solve this problems banks tried to integrate their constructed delivery channel. But they got another problems integrating delivery channel constructed by other kind of technology. Need delivery channel system integrating every client connections to a single solution and by using this techniques they can collect clients connection history more effectively is what banks reached. Gartner Group, March 2000, mentioned integrate of delivery channel system in report 'Retail Banking'. This report added diversify Back-end system and integrate connection to integrate delivery channel system deducted in Tower Group seminar. After 90s, by Down Sizing Back-end system are getting diversification. But established delivery channel system are maintaining to connection of a single Back-end system (HOST system). By this system, they needed a system connection between Back-end systems. And this causes complicating system and hard to achieve integrity. To solve these side effects, Gartner Group suggested that Delivery Channel System should be possible to be connected to various Back-end systems. Based on Tower Group's and Gartner group's announcements, pivoting on companies providing terminals to business branches established concepts and developed Delivery Channel Solution (DCS). This system focused on interfaces accessing various back-end systems simultaneously, interfaces connected to various customers' point of contact such as a terminal system in branches, call center, internet banking, and so on.

However to appeared various channel continuously, especially, utilization of the channel is diversifying by using the Internet. It need to change system of combining delivery channel system conceptually [4]. To accommodate these changes, Delivery Channel System (DCS) in a single concept of MCA, which provides a channel specify server and common function that handles the unique features of each contacting customer system is differentiated to MCA (Multi Channel Architecture). But there is a difference in the realization process, MCA is divided into DCS conceptually. Meaning and utility of MCA can be considered the same as the DCS.

2 Related Work

MCA is part of united Delivery Channel System. And we can call it as a set of common function (except indigenous functions of the server) in united Delivery Channel System [3]. Concepts of MCA's structures and functions are shown at Fig. 1.

FII (Front-end Interface Integration) provides function which distributes results of process to a indigenous server of channel requesting trade with interface which is easy to connect channel's indigenous server. ACR (Access Control Rule) provides expanded management functions of restricting deals according to users and whether it is possible to deal in each channel. MFR (Message Formatting Rule) recasts full text of dealing request into a type that can communicate with Back-end system. In case, combination deal, can be built into several text fits to Back-end system. After deal it converts and combines to suitable type for channel's indigenous server. MFR reforms the deal requesting script into appropriate form in order to transfer to it's back-end system. In complex dealing cases, more than one script and be formed for each appropriate Back-end system. TCR (Transaction Control Rule) decides transforming rules that sends deals to Back-end system, managing deals sent to numberous Back-ends and process of canceled deals. LTR (Log Trigger Rule) restricts level of log to collect client access history, finally connects to diversified Back-end system through BII (Back-end Interface Integration). Adding these kind of services, IRA (Integrates Rule Authoring) which makes easy to define rules for effectiveness of developing and operating, and IAC (Integrated Administration Center) which is for supervising the system are provided.

By increasing frequency of using smart device, number of devices which wants to connect to financial servers has increased. By lack of manpower for updating MCA server to confront new devices, needs of MRM technology which can respond actively to newly added devices are increasing. Existing technology have problems such as follows.

Functional problem: Existing tools are hard to define meta field data, and hard to upgrade or modify the tool to newly added devices. Economical problem: Developing for every and each new devices causes waste of TCO (Total Cost of

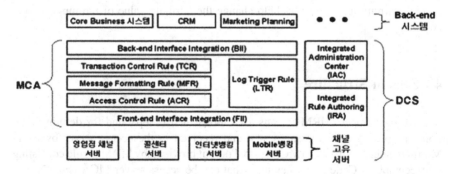

Fig. 1 Conceptual structure of MCA

Ownership) such as time and money. Reliability problem: Can't cope with every outbreak situation, because of different of developer for each mapping tools of devices. Efficiency problem: Coding for MCA server causes waste of manpower. Efficiency problem: Coding for MCA server causes waste of manpower.

3 Requirement Analysis

3.1 Function-Related Requirement

Make MCA can handle actively with every new devices. Type of meta data can easily modified. MRM's final output result is Target-language which can run at MCA. Can process 1 Transaction for each connected device. Make UI have easy Rule Mapping tool.

3.2 Interface-Related Requirement

When rule mapping occurs, it is necessary to build a user interface simply and efficiently to deal with all terminals' messages actively. In addition, for better performance and stability, C language or Assembler language can be an option. Especially in C language, C# language is better for providing familiar Window forms.

3.3 User-Related Requirement

In order to manage data required by the back server easily, it makes easy to convert a meta field data and be able to generate target code. To make it easier to change the data type and attribute values in the bank, use the Window Form which don't need special learning how to use to help change the attribute value of the input, output and data.

4 System Structure

MCA Rule Mapper (MRM) means a UI tool which is separated for defining relationships between the messages sent and received in the access server by using Multi-channel architecture (MCA). Final output provided by MRM is rule consisting of executable C Code target language that can be access server MCA (Fig. 2).

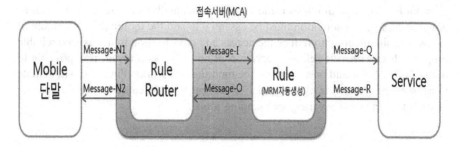

Fig. 2 System configuration; *Message—N1* input message from Mobile terminals (Header + SaveKey + Data); *Message—N2* ouput message from Mobile terminlas (Header + SaveKey + Data); *Message—I* message from terminals (Header + SaveKey + Data); *Message—O* response message to *Message—I* (Header + SaveKey + Data); *Message—Q* input message to service (Header + Data); *Message—R* output message from service (Header + Data)

5 Module Specification

MRM is largely divided into three parts; meta-field editor, message editor, and message mapper, hence it takes three steps.

Meta-field editor—Meta-field editor is used to define fields in various message types; XML format is used to define meta-field such as No., English name, Korean name, data type, and data length, and saved as a file named MESSAGENAME. meta. Meta-field editor can provide several functions to generating fields in a file.

Message editor—To generate message form, fields defined by meta-filed file can be used. This message, then, consists of fields that can contain information derived from an input message or service message. Message is saved as a file named "MESSAGENAME.msg" with XML format. Message editor provides grid input interface showing meta-fields, so that fields can be selected and inserted from grid. A message file can handle only one message.

Message mapper—Message mapper makes rule code mapping from input message to service message or vice versa. Once a channel input message and a service message are defined, Message mapper matches corresponding fields of each message. When a mapping occurs, there should be defining properties of mapped fields. It should provide a sim.

6 Anticipated Results and Contributions

When you add a new terminal other than terminal which is being used in the part of MCA processing the messages sent by the terminal, you need to modify default program. By analyzing information of messages sent from new terminals, MRM generates *.c and *.h files about the terminal. If you set up a link between these two files, you can add terminals without adjusting programs. Because more of the terminal is very large in number, the way to fix manually is not feasible, as well as not visible. So the automation must be essential to add the terminal. Furthermore, it is possible significantly to reduce the access and modification to the existed reliable program. It can provide reliable service continuously. As a result, it is possible to minimize the labor and cost spent on the modification works. Finally it will be able to achieve the standardization in order to prevent the loss caused by the difference of each tool in using a variety of mapping tools.

References

1. Carignani A (2000) Supporting a multiple channel architecture design: the UML contribution in a virtual banking environment. Universita Catolica del Sacro
2. Pavlovski C (2013) A multi-channel system architecture for banking. IBM
3. Sun J (2010) Building a common enterprise technical architecture for an Universal Bank. Dalian Maritime Univ
4. Pousttchi K (2004) Assessment of today's mobile banking applications from the view of customer requirements. Augsburg Univ

A Simple Fatigue Condition Detection Method by using Heart Rate Variability Analysis

U.-Seok Choi, Kyoung-Ju Kim, Sang-Seo Lee, Kyoung-Sup Kim and Juntae Kim

Abstract The traffic accident statistics show that fatigue (drowsiness) and drunk driving are the major causes of traffic accidents. Therefore, it is important to detect and prevent driving in fatigue condition. The conventional fatigue detection technologies use methods that detect a driver's drowsiness from the direction of the face, the eye closing speed, etc., using cameras and various senses. Such technologies, however, are not only expensive but also have positional detection limitations as cameras and sensors are used, thereby restricting the driver's behavior. In this study, a simple method of detecting fatigue condition based on HRV (Heart Rate Variability) data is presented. The proposed method can greatly reduce the cost of drowsiness prevention system for safe driving.

Keywords HRV · PPG · Fatigue · Drowsiness analysis

1 Introduction

Fatigue is the regular physiological phenomena of the temporary decrease of working ability due to long time work. The drivers lose their willpower instantaneously when they are in a state of sudden fatigue, so it has led to a serious traffic

U.-S. Choi · K.-J. Kim · S.-S. Lee · K.-S. Kim · J. Kim (✉)
Department of Computer Science and Engineering, Dongguk University,
Seoul, South Korea
e-mail: jkim@dongguk.edu

U.-S. Choi
e-mail: cws1340@gmail.com

K.-J. Kim
e-mail: vkcjs017@gmail.com

S.-S. Lee
e-mail: sangseo13@naver.com

K.-S. Kim
e-mail: kyskks@dongguk.edu

© Springer Science+Business Media Singapore 2016
J.J.(Jong Hyuk) Park et al. (eds.), *Advances in Parallel and Distributed Computing and Ubiquitous Services*, Lecture Notes in Electrical Engineering 368,
DOI 10.1007/978-981-10-0068-3_27

accident. According to statistics, the traffic accidents of driving of fatigue accounted for about 70 % in the total of traffic accidents. If we can develop a fatigue detection algorithm, it can be used to give alert to drivers before going into sleep to prevent accidents.

The previous technologies of preventing accidents used various expensive devices, such as cameras to sense the pupil state of eyes or the face direction, seat pressure sensors and vibration devices, etc. Those devices are usually had problems of high cost and low accuracy. Another way of detecting fatigue condition is to use heart beat changes by using heart rate monitor. In this paper we propose a method of detecting fatigue condition based on HRV (Heart Rate Variability) data. HRV devices are cheaper compared to other devices, but can closely analyze heartbeats and can be used to judge users' fatigue state more easily.

2 Related Work

Heart rate variability is defined as the measure of variation in heart beats. It is calculated by analyzing the beat to beat intervals (the R–R intervals), and it provides a passive means to quantify drowsiness [1]. The HRV has been used to examine mental workload and stress, and a change in HRV can be an indication of decrease in mental workload, which can occur in sleepy drivers over long driving [2, 3].

The method of measuring HRV is divided into two parts which are time domain analysis and frequency domain analysis. Time domain analysis are common and simple to perform. It usually calculate the standard deviation of heart beat intervals. Frequency domain analysis is based on mathematical transformations of the signals from time domain to frequency domain. Power spectral density (PSD) analysis provides the information of how power distributes as a function of frequency [4].

3 Data Collection and Analysis

3.1 Time Domain Analysis

To detect fatigue condition, we use the HRV values and the standard deviation of RR. R is a point corresponding to the peak of the ECG (Electrocardiography) wave, and RR is the interval between successive Rs. The measurements we propose for fatigue detection are as follows:

SDNN: It is the standard deviation of normal R-R.

$$SDNN = \sqrt{\frac{\sum_{i=1}^{N} \left(RR_i - \overline{RR}\right)^2}{N}}$$

rMSSD: It is the root mean square of the squares of the differences between neighboring R-R. It calculates the changes between neighboring R-R and then responds on rapid change of HRV.

$$rMSSD = \sqrt{\frac{\sum_{i=1}^{N}(RR_i - RR_{i-1})}{N-1}}$$

3.2 HRV Data of PPG Measuring Device

The PPG (photoplethysmogram) device is used to measure HRV. PPG is an optically obtained volume measurement of an organ. With each cardiac cycle the heart pumps blood to the periphery, and the change in volume caused by the pressure pulse is detected by illuminating the skin with the light from a light-emitting diode (LED) and then measuring the amount of light either transmitted or reflected to a photodiode. Each cardiac cycle appears as a peak. We collected PPG dataset of about 10 people for normal (awake) state and sleepy state, then convert them to HRV values and analyze those values. We measured specific people several times to get reliable HRV data.

By analyzing PPG graph slopes, we can detect the peak points, and convert them to HRV by measuring the time between adjacent peaks. Figure 1 shows the peak points of PPG and Fig. 2 shows the HRV graph converted from the PPG.

The PPG data is converted to HRV, and then the time domain analysis is performed to compute various measures for fatigue detection. Figure 3 shows these procedure of fatigue detection.

Fig. 1 Time between peak points of PPG

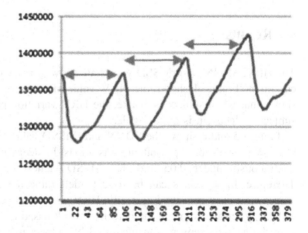

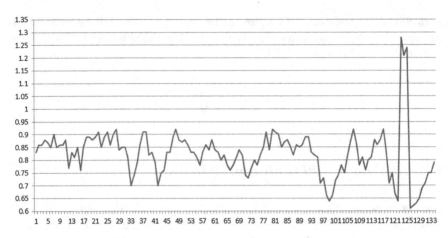

Fig. 2 A sample HRV of normal state (awake)

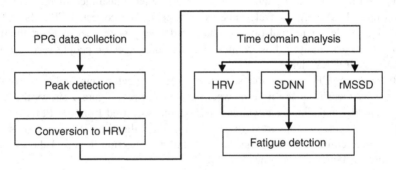

Fig. 3 The procedure of fatigue detection from PPG

4 Results

The HRV, SDNN, and rMSSD values are shown to be closely related to the fatigue condition. Figure 4 shows the HRV values for awake state and sleepy state. It shows that when subject is awake, the HRV variation pitch irregularly, but when subject is sleepy, it is quite regular.

Figures 5 and 6 shows the SDNN and rMSSD values for awake state and sleepy state. What we should pay attention is y axis. In sleepy state, the SDNN values are continuously under 0.05 and the rMSSD values are continuously under 0.1. Therefore, these values can be used to determine the fatigue condition. If PPG values have been measured correctly, HRV values were calculated in error range of ± 0.03. When we compare the detection result based on SDNN and rMSSD graph with the actual sleepy region measured by subject using a timer, it shows higher than 90 % accuracy. Table 1 shows the results.

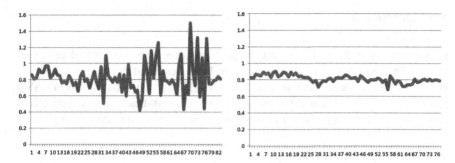

Fig. 4 HRV values for (L) awake, (R) sleeping

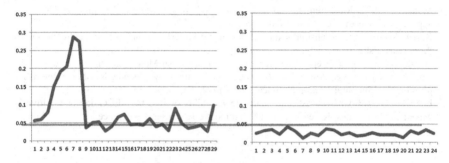

Fig. 5 SDNN values for (L) awake, (R) sleeping

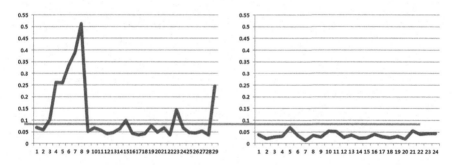

Fig. 6 rMSSD values for (L) awake, (R) sleeping

Table 1 Experiment results

	HRV (%)	SDNN (%)	rMSSD (%)
In awake state	94	93	93
In sleeping state	96	94	93

5 Conclusion

In this study a method of detecting fatigue condition based on HRV data is presented. The PPG device is used to measure heart beats and they are converted to HRV values. To detect fatigue condition the HRV graphs and SDNN, rMSSD measures are used which represents changes of heart beats. The experimental results shows that the proposed method can detect fatigue conditions easily.

References

1. Mulder L (1992) Measurement and analysis methods of heart rate and respiration for use in applied environments. Biological Psychology
2. Bornas X et al (2004) Self-implication and heart rate variability during simulated exposure to flight-related stimuli Anxiety, Stress, and Coping
3. Horne A et al (1995) Sleep related vehicle accidents. British Medical Journal
4. American Heart Association (1996) Heart rate variability: Standards of measurement, physiological interpretation, and clinical use. Circulation
5. Lee E et al (2013) Characteristics of heart rate variability derived from ECG during the driver's wake and sleep states. Korea Society of Automotive Engineers Conference

Insider Detection by Analyzing Process Behaviors of File Access

Xiaobin Wang, Yongjun Wang, Qiang Liu, Yonglin Sun
and Peidai Xie

Abstract Information security is a great challenge for most organizations in today's information world, especially the insider problem. With the help of malwares, insiders can search and steal valuable files easily and safely in an organization's network. In this paper, we collect a dataset of file access behaviors for normal processes and malware processes. We analyze the dataset and find several features in which normal processes and malware processes show significant differences, a file access behavior model is given based on these features, and we apply both semi-supervised and unsupervised approaches to verify the effectiveness of our model, experimental results demonstrate that our model is effective in distinguishing between file access behaviors of normal processes and malware processes.

Keywords Information security · Insider · Insider detection · Malware · File access behaviors

X. Wang (✉) · Y. Wang · Q. Liu · Y. Sun · P. Xie
College of Computer, National University of Defense Technology, Changsha, China
e-mail: wxb1107@hotmail.com

Y. Wang
e-mail: wwyyjj1971@126.com

Q. Liu
e-mail: libra6032009@gmail.com

Y. Sun
e-mail: leomuyi@126.com

P. Xie
e-mail: peidaixie@gmail.com

© Springer Science+Business Media Singapore 2016   209
J.J.(Jong Hyuk) Park et al. (eds.), *Advances in Parallel and Distributed Computing and Ubiquitous Services*, Lecture Notes in Electrical Engineering 368,
DOI 10.1007/978-981-10-0068-3_28

1 Introduction

Most organizations today relay upon computers and networks to deal with all kinds of information data, which brings them huge benefits. However, the inherent defects of computers and networks make the risk of information security increase. Although several security facilities, e.g. Firewall and Intrusion Detection System (IDS), have been deployed to protect information from outer attacks, they are weak to defeat insiders [1–4].

Several approaches have been proposed to solve the problem of insider detection [5–7], none focuses on the case that insiders use malwares to search and steal files. We argue that using malware to search files is a preferable way for insiders. On one hand, using malware is a more effective and safer way for insiders; on the other hand, intranet users (especially those who are physically separated in the Internet) do not often pay much attention on the security of the intranet itself. Therefore, malicious insiders can easily inject malwares into other computers in the intranet.

In this paper, we extend the work of [7], and propose a novel approach to detect malicious insiders who use malwares to search files. We collected a real dataset of process file access behaviors, including behaviors of normal processes and those of Trojan processes that were remotely controlled by an attacker to search files located in the target computer. By analyzing the dataset, we found out several important features of file access behaviors, which were effective to discriminate malware processes from normal ones. Then, we built a file access behavior model based on these features, and we applied the OCSVM (One Class Support Vector Machine) and the k-means algorithm to verify the effectiveness of our model in detecting abnormal file access behaviors.

Organization. The rest of this paper is organized as follows. In Sect. 2, we discuss related works on insider detection. In Sect. 3, we introduce the dataset of file access behaviors, followed by preliminary analysis results, as well as the file access behavior model. Next, in Sect. 4, we apply both semi-supervised and unsupervised algorithms to verify the effectiveness of our model in detecting abnormal file access behaviors. Finally, Sect. 5 concludes the paper and presents the future works.

2 Related Work

Behavior profiling is a common way in insider detection, but the behavior data we need to detect insiders are different according to different application areas, including command line calls and file access behaviors, etc.

Previous researches of insider detection focused on profiling user behaviors by the command line calls they issued, the SEA (Schonlau et al.) dataset [6] is a de facto standard dataset in this area. Several approaches [8, 9] have been proposed based on this dataset, however, the results they got are far from good. Followers either tried to improve the detection algorithm [10–13] or enrich the dataset [14, 15], though better results they got, still not good enough.

Prior approaches in the literature that have the most similarity to ours is [7], in which M. Ben-Salem models user search behaviors for masquerader detection, the approach of their work focus on the search behavior of users while our work focus on those of processes. The features they choose only characterize search volume and velocity, which is not enough to achieve a high detection rate of malware process detection. Besides search volume and velocity, our work also choose other features of file access behaviors, including file path dispersion, operation type, and file type, in which normal processes and malware processes show significant differences.

3 File Access Behavior Analysis

In this section, we first introduce the dataset of file access behaviors. Then, we analyze the dataset in four aspects: file path dispersion, operation type, file type and velocity.

3.1 Dataset

In order to collect a dataset of file access behaviors, we have developed a file access monitor which runs on the file system driver level for Windows platforms, the monitor records all the file access events of processes running under the operation system. Examples of some file access records are given in Table 1.

- File access records of normal processes

We reinstalled the operation system before collecting file access records of normal processes, so that we can ensure there is no malware process running when collecting normal data. The data collected spanned 7 days and 5 h every day on average, during the collecting period, the user took routine operations on the computer, e.g. editing some documents, programming using Eclipse and Visual Studio, etc.

Table 1 Examples of file access records

Time	Process name	Process id	Operation type	File path
18:16:43.581	services.exe	520	CloseFile	C:\Windows\System32 \services.exe
19:26:23.477	svchost.exe	924	QueryDirectory	C:\WINDOWS\system32
19:41:07.641	WINWORD. EXE	2380	ReadFile	G:\study\paper \experiments.docx

The normal data takes about 500 Mbytes and 4 million records in total, and the number of processes is 37.

- File access records of malware processes

After collecting the normal data, we installed 17 Trojans on the target computer, all these Trojans can be downloaded from Internet. We then asked 30 students in our lab to operate on the control side of the Trojans (some students use a same Trojan), the students knew nothing about the target computer and they were told to search sensitive files on it through the remote control capability of the Trojans, so that their behaviors would be very similar to those of malicious insiders who try to search and steal valuable files on other ones' computers. We collected the process file access records when they were using the Trojans to search files, each student operated 30 min on average, these file access records belong to the suspicious data, including file access records of both normal process and Trojan processes, it takes about 240 Mbytes and 2 million records, with 17 Trojan processes and others are normal processes.

3.2 File Access Behavior Analysis

Based on the dataset collected, we analyzed several features of process file access behaviors, including file path dispersion, operation type, file type and velocity, we found that normal processes and the malware processes show significant differences in some of these features. We divided file access records into groups according to a time interval of 20 s, the values of the above features were computed for each group of each process, and we then compared these groups of features values between normal processes and malware processes.

3.2.1 File Path Dispersion

The distribution of files accessed by a process is decided by the capabilities of the process. Usually, a normal process only needs to access certain files to accomplish its capabilities, including system files and the process related files. However, a malware process controlled by an insider will probably access files widely spread the file system of the target computer, aims for searching valuable files as many as possible.

In order to quantifiably represent the distribution of files accessed by a process, we proposed an algorithm of computing FPD (File Path Dispersion) [16], which is a quantized value that measures how far a set of file paths is spread out. We use a Variance-like formula to compute FPD values, for a given set of file paths $P = \{p1, p2, \ldots pn\}$, FPD is given by

$$FPD = \frac{1}{n}\sum_{i=1}^{n}(p_i - \bar{p})^2.$$

As shown in Fig. 1, we computed FPD values for every group of files accessed by normal processes and malware processes, <10 % FPD values of malware processes are lower than 0.1, far fewer than normal processes. According to the result of Fig. 1, we can infer that:

- Malware processes are not always access files widely, they behave like normal ones when they are not executing file searching capabilities.
- Normal processes will access files widely sometimes, for example, when Antivirus is scanning disks.

3.2.2   Operation Type

Having analyzed the FPD feature of our data, we now analyze the operation type and frequency of file access behaviors. As shown in Table 2, we organize operation types into 4 categories:

- Create/Close: When a process start to access a file, it will carry out a "CreateFile" operation to the file system, and after finishing accessing the file, it does a "CloseFile" operataion.
- Read/Write: a process will carry out "ReadFile" or "WriteFile" operation when it need to read or write to a file.
- Query: there are several Query operations supported by the file system for processes to query special information of a file/directory, the most common ones are "QueryOpen", "QueryDirectory" and "QueryXXXInformationFile" (e.g. QueryNameInformationFile et al.).
- Others: we classify all other operation types into this category.

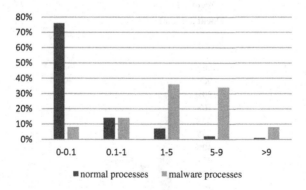

Fig. 1 Distribution of FPD values

Table 2 Distribution of different operation types

Category	Operation type	Malware processes (%)	Normal processes (%)
Create/Close	CreateFile	18	14
	CloseFile	17	13
Read/Write	ReadFile	5	25
	WriteFile	4	5
Query	QueryOpen	3	15
	QueryDirectory	41	4
	QueryXXXInformationFile	4	13
Others	Other operation type	8	11

3.2.3 File Type

Besides operation type differences, types of accessed files are also different between malware processes and normal processes. As shown in Table 3, malware processes access more directories than normal processes (53 % compared to 14 %), as well as accessing of documents (including txt, doc, pdf, ppx, et al.). But the total count of different file types are similar between malware processes and normal processes, this means that both of them access a wide range of different types of files.

3.2.4 Velocity

Figure 2 shows that nearly 90 % time that normal processes do not access any file, while the corresponding percentage for malware processes is a little lower (71 %). Except the 0-velocity parts, malware processes and normal processes have similar

Table 3 File type differences

File type	Malware processes	Normal processes
Count of file types	98	103
Directory (%)	53	14
Document (%)	4	1

Fig. 2 Velocities distribution

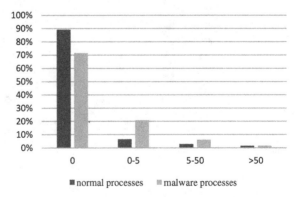

■ normal processes   ■ malware processes

Fig. 3 Velocities distribution (non-zero part)

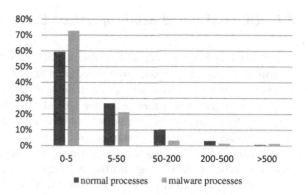

file access velocities, as shown in Fig. 3. This overthrows one of our viewpoints that we used to think malware processes would access files faster than normal processes.

3.3 File Access Behavior Model

The above analysis shows that malware processes have very different characteristics compared to normal processes. Based on these results, we build a file access behavior model which can be used to distinguish between malware processes and normal processes. The model is represented as a Vector S = <f1, f2, ... f13>, which contains 13 features according to the above analysis.

- The FPD feature: The FPD value for each group of process file access records is an important feature of the model.
- Operation type features: As mentioned in Sect. 3.2.3, there are 8 kinds of operation types, and we use the percentage of each operation type as a feature.
- File type features: We leverage 3 file type features: total count of different file types, the percentage of the directory type, and the percentage of the document types.
- The velocity feature.

4 Model Verification

Based on the file access behavior model described in Sect. 3.3, we launched 2 kinds of experiment to verify the effectiveness of detecting malware processes used by insiders to search and steal files in target computers, including OCSVM based semi-supervised detection, and k-means based unsupervised detection.

4.1 OCSVM Based Semi-supervised Detection

The One Class Support Vector Machine (OCSVM) is an extension of the SVM algorithm, it was first proposed by Scholkopf et al. [17]. OCSVM can be viewed as a regular two-class SVM where all the training data lies in the first class and the origin is taken as the only member of the second class. In this paper, we use libsvm [18] to perform our experiment, the svm_type we choose is one-class SVM, and we use a RBF kernel, other parameters are set to default values.

We launch the experiment several times, an average classification result is shown in Table 4, here, false positive of groups means that groups of file access records belong to normal processes were classified as abnormal, and false negative of groups means that groups belong to abnormal processes were classified as normal. A process is identified as abnormal when the count of its file access record groups that were classified as abnormal surpasses a certain value n. We tested varying n from 1 to 90 (every process in testing data spans 30 min, with 90 groups of records according a time interval of 20 s), and get a best result when n is between 7 and 11, with a ≤ 2.7 % false positive and a ≤ 6.6 false negative. The false positive is relatively high if n is smaller than 7, and when n is bigger than 11, the false negative increase greatly.

4.2 K-Means Based Unsupervised Detection

While our OCSVM results are quite good, an OCSVM based approach is still a semi-supervised learning tool, it is of limited use in practice: it requires training on large samples of training data. In this section, we propose an unsupervised detection approach based on the k-means [19] clustering algorithm, which require no training data. As the k-means algorithm aims to partition n observations into k clusters, it can be used to partition our file access data into different clusters. In this experiment, we combined the normal data and the suspicious data together as a testing dataset, and we set k = 2 because there are only two clusters: normal and abnormal.

As described in [16], the computing of FPD values rely upon training data to get the most frequently accessed files and directories $\bar{p}$. In order to compute FPD values in the unsupervised approach, we randomly select several file paths in a group of

Table 4 Result of the OCSVM detection

Value of n	False positive of groups (%)	False negative of groups (%)	False positive of processes (%)	False negative of processes (%)
n = 2	5.9	42.8	18.9	0
n = 7	5.9	42.8	2.7	3.3
n = 11	5.9	42.8	0	6.6
n = 20	5.9	42.8	0	36.7

Table 5 Result of the unsupervised detection based on k-means

Value of n	False positive of groups (%)	False negative of groups (%)	False positive of processes (%)	False negative of processes (%)
n = 2	10.4	29.7	73	0
n = 7	10.4	29.7	18.9	0
n = 11	10.4	29.7	5.4	6.6
n = 15	10.4	29.7	2.7	9.9
n = 20	10.4	29.7	0	36.7

file access records to serve as $\bar{p}$. From a statistical point, the randomly selected file paths are still in accordance with $\bar{p}$, so the FPD values of normal processes will not change greatly, and for malware processes, the FPD values will still remain relatively large. Other 12 features of the file access behavior model can all be computed without training data, they remain the same as the OCSVM approach.

Table 5 shows the classification result of the unsupervised approach, the false positive of groups is bigger than that of the semi-supervised approach while the false negative is smaller. The best detection result is presented when n is between 11 and 15, bigger than the OCSVM approach, with a false negative < 5.4 %, and a false negative < 9.9 %, though a little worse than the OCSVM approach, it is still effective.

4.3 Discussion

Although both semi-supervised and unsupervised detection experiments demonstrate that our file access behavior model is effective in detecting malware processes, it is still not real time verifications. In order to verify the effectiveness in real time detection, we applied these two approaches to a prototype system which is used to detect abnormal file access behaviors of malware processes.

We installed the prototype system to a computer in our lab, as well as several Trojans. Most file searching activities of Trojans were detected in less than 1 min. However, there are 2 normal processes which were sometimes mistakenly identified as abnormal processes.

- One is the "Explorer.exe", most UI activities are executed by this process, including manually file search activities through the explorer of the operation system. Users may search and access different files in different times, and these activities are all executed by "Explorer.exe", that is the reason why our prototype system sometimes make incorrect detection decisions.
- The other process is a desktop file searching process, it is identified as abnormal when we use this process to search documents.

The above problem reveals that our prototype system still have some limitations. In our future works, we will try to improve our file access model as well as the prototype system, in order to solve the false alarm problem.

5 Conclusion and Future Work

In this paper, we collect a dataset of file access behaviors for normal processes and malware processes. To the best of our knowledge, we are the first to analyze the file access behaviors of processes. We analyze the dataset and find several features in which normal processes and malware processes show significant differences. Based on these features, we build a file access behavior model, and we apply an OCSVM based semi-supervised approach and a k-means based unsupervised approach to verify the effectiveness of our model in detecting abnormal file access behaviors. Experimental results demonstrate that our model is effective in distinguishing between normal processes and malware processes.

We believe file access behavior models of processes can be an effective technology for process behavior profiling in detecting malware processes. One of our future works is to improve our file access behavior model with other file access related features, and enhance the detection efficiency. We will also study the network behavior when insiders transfer files in a network, so that we can not only detect which computer is attacked by insiders, but we can also detect where the insider is located in a network.

Acknowledgements This paper is supported by the National Natural Science Foundation of China (Grant No. 61271252 No. 61202482), and the Open Fund Project of Innovation Platform for Universities in Hunan Province (Grant No. 13K025).

References

1. Bishop M, Gates C (2008) Defining the insider threat. In: CSIIRW'08
2. PfLeeger SL, StoLfo SJ (2009) Addressing the insider threat. In: IEEE security and privacy Nov/Dec, pp 22–29
3. Theoharidou M, Kokolakis S, Karyda M, Kiountouzis E (2005) The insider threat to information systems and the effectiveness of ISO17799. Comput Secur
4. Blackwell C (2009) A security architecture to protect against the insider threat from damage, fraud and theft. In: CSIIRW '09
5. Salem MB, Hershkop S, Stolfo SJ (2008) A survey of insider attack detection research, in insider attack and cyber security: beyond the hacker. Springer, New York
6. Schonlau M et al (2001) Computer intrusion: detecting masquerades. Stat Sci 16:1–17
7. Salem MB, Stolfo SJ (2011) Modeling user search behavior for masquerade detection. In: Proceedings of the 14th international symposium on recent advances in intrusion detection. Springer, Heidelberg, pp 1–20

8. Maxion RA, Townsend TN (2002) Masquerade detection using truncated command lines. In: Proceedings of the international conference on dependable systems and networks, pp 219–228
9. Coull S, Branch J, Szymanski B, Breimer E (2003) Intrusion detection: a bioinformatics approach. In: Proceedings of the 19th annual computer security applications conference
10. Yung KH (2004) Using self-consistent Naïve-Bayes to detect masqueraders. Stanford Electr Eng Comput Sci Res J (2004)
11. Oka M, Oyama Y, Abeand H, Kato K (2004) Anomaly detection using layered networks based on Eigen co-occurrence matrix, RAID, pp 223–237
12. Wang K, Stolfo S (2003) One-class training for masquerade detection. In: Proceedings of the 3rd IEEE conference data mining workshop on data mining for computer security November, pp 10–19
13. Seo J, Cha S (2007) Masquerade detection based on SVM and sequence-based user commands profile. In: ASIACCS'07, 20–22 March 2007, pp 398–400
14. Greenberg S (1988) Using Unix: collected traces of 168 users. Technical report 88/333/45, Department of Computer Science, University of Calgary, Calgary, Canada
15. Maxion RA (2003) Masquerade detection using enriched command lines. In: Proceedings of the international conference on dependable systems and networks (2003)
16. Xiaobin W, Yonglin S, Yongjun W (2014) An abnormal file access behavior detection approach based on file path diversity. In: Proceedings of the international conference on information and communications technologies, pp 455–459
17. Scholkopf B, Platt J, Shawe-Taylor J, Smola A, Williamson R (2001) Estimating the support of a high-dimensional distribution. Neural Comput 13(7):1443–1472
18. Chang C-C, Lin C-J (2011) LIBSVM: a library for support vector machines. ACM Trans Intell Syst Technol 2(27):1–27
19. Anil KJ (2010) Data clustering: 50 years beyond K-Means. Pattern Recogn Lett 31(8):651–666

Analysis of the HOG Parameter Effect on the Performance of Vision-Based Vehicle Detection by Support Vector Machine Classifier

Kang Yi, Seok-Il Oh and Kyeong-Hoon Jung

Abstract Support Vector Machine (SVM) classifier with Histogram of Orientated Gradients (HOG) feature is one of the most popular techniques used for vehicle detection in recent years. In this paper, we study the effect of HOG parameter values on the performance and computing time of vehicle detection. The aim of this paper is to explore the relationship between performance/computing time and HOG parameter values, and eventually to guide finding the most appropriate parameter set to meet specific problem constrains.

1 Introduction

The demands for Advanced Driver Assistance Systems (ADAS) are constantly increasing in recent years for safety, convenience, and comforts of vehicle driving. One of the most common ADAS functions is vision-based vehicle detection. The detection results are utilized for further enhanced high-level functions such as Forward-Collision Warning (FCW) to prevent fatal accidents. A typical scheme for vehicle detection uses Histogram of Oriented Gradients (HOG) coupled with

This work was supported by the Center for Integrated Smart Sensors funded by the Ministry of Science, ICT and Future Planning as "Global Frontier Project" (CISS-2011-0031863).

K. Yi (✉) · S.-I. Oh
School of Computer Science and Electrical Engineering, Handong Global University, Pohang, Republic of Korea
e-mail: yk@handong.edu

S.-I. Oh
e-mail: galsian@naver.com

K.-H. Jung
School of Electrical Engineering, Kookmin University, Seoul, Republic of Korea
e-mail: khjung@kookmin.ac.kr

© Springer Science+Business Media Singapore 2016
J.J.(Jong Hyuk) Park et al. (eds.), *Advances in Parallel and Distributed Computing and Ubiquitous Services*, Lecture Notes in Electrical Engineering 368,
DOI 10.1007/978-981-10-0068-3_29

Support Vector Machine (SVM) classifier [1]. HOG is a robust object feature that represents edge features properly and resists illumination changes. The SVM is very powerful tool for data classification in which the classification is achieved by a separating surface called hyper-plane in the input space of the dataset [2].

In general, there is a trade-off relationship between the accuracy and the computing time of vehicle detection by HOG with SVM. It is certain that the HOG descriptor structure affects the detection accuracy and computing time. And the HOG parameters such as window size, block size, cell size, and bin size eventually determine the structure of HOG descriptors [3]. However, the effects of HOG parameter are closely related each other and the combination of parameters affects the final performance and computing time. Therefore, what we need to find is not a value of single HOG parameter but a set of values of HOG parameters.

Based on our experience, the structure and dimension of feature vector, which are defined by the HOG parameters, have strong influences on the classification results by SVM in terms of accuracy and computing time. In this paper, we examine the parameter space of HOG features by exhaustive experiments to find the optimal combination of HOG parameters. And, we suggest a way to determine the parameter set which meets a specific problem constrains and objectives.

2 HOG Descriptor Parameters

2.1 Computation Steps of HOG Features

The HOG, which represents the features of an object by a set of orientations of gradients of image pixel intensities, can be obtained by the following steps. The first step of HOG computation is to divide the given image window by the given equal sized squares named cell. The second step is to create a histogram for each cell. Every pixel within a cell cast a vote for orientation-based histogram bins which consist of evenly spaced angles between 0° to 180° for unsigned gradient or 0°–360° for signed gradients. In this paper, for vehicle detection, we use unsigned gradients in the range 0°–360°. Gradient for each pixel may be computed by edge detection operators such as Sobel filter. The third step is a gradients normalization step. In order to account for the resulting orientation changes due to the illumination and contrast differences, the magnitude of gradients need to be normalized by considering neighboring cell gradients values. The cell normalization is performed over a group of cells named block. Each block consists of connected cell forming an N × N rectangles. Each of the blocks are typically overlapped by a multiple of cell size, the overlapping size is called stride size. With non-zero stride size each cell contributes more than once for the final HOG descriptor. The final step is to construct the HOG descriptor which is the vector of normalized cell histograms from all blocks in an image window.

Fig. 1 The structure of cell and block for HOG descriptor

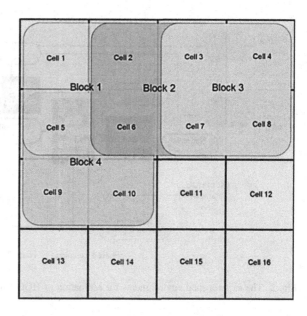

2.2 Parameters of HOG Descriptor

The HOG parameters mentioned above are the window size, block size, stride size, cell size, and number of bins for each cell. The window size, block size, and stride size should be multiple of cell size. Figure 1 illustrates the HOG structure in which the block size is double of cell size and the stride size is the same as a cell size. The HOG descriptor dimension for a window is calculated by multiplying the number of cells per block and the number of blocks per window as shown in Eq. (1).

$$HOG\_dimension = BINS \times [BS/CS]^2 \times [(WS-BS)/SZ + 1]^2 \qquad (1)$$

where BINS denotes the bin size (the number of bins in a histogram) and WS, BS, CS, SZ represents the window size, block size, cell size, and stride size, respectively.

3 HOG Parameter Space Exploration by Experiments

3.1 Experimental Setup

The overall experimental procedures and data flow are shown in Fig. 2. For each sample in the training data set, the HOG feature is extracted, and then the features and their corresponding labels, which are marked as one of four positive categories

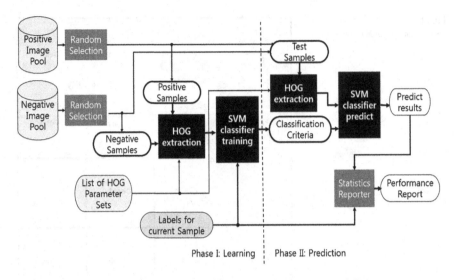

Fig. 2 The experimental environments for estimation of HOG parameter influence

by vehicle poses or negative category, are provided for the supervised machine learning by SVM. As a result of phase I, the classification criteria are obtained. In phase II, the classification criteria from the previous phase and the test samples are fed to SVM classifier to obtain statistics for current HOG parameter set. For each HOG parameter value set the mentioned testing procedures are iterated several times to get reliable performance evaluation data.

We trained the SVM with a set of positive images with various vehicle poses as well as negative images. The considered vehicle poses are shown in Fig. 3. Each positive class sample pool consists of more than 1,000 cropped images from vehicular black box rear camera. About 90 % of the samples from the pool are used for training and 10 % is used for testing. The testing and training samples are randomly selected.

3.2 Experimental Results

In the experiments, we first figured out the most proper bin size of HOG. Based on our pre-experiments the eight is the best value of bin size for almost cases. In this paper, we explore the relationship between HOG parameter values and computing time or detection accuracy under the fixed bin size (BINS = 8).

We applied various parameter sets with fixed BINS = 8 as shown in the left part of Table 1. We restrict the parameter values range as WS > BS > CS = SS to prune out the parameter value search space efficiently.

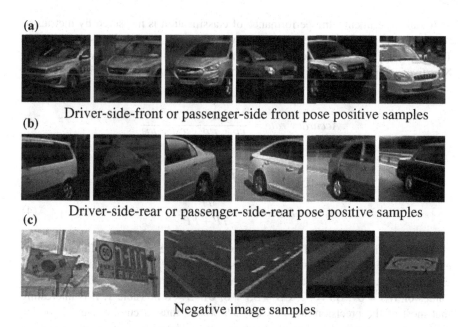

(a) Driver-side-front or passenger-side front pose positive samples

(b) Driver-side-rear or passenger-side-rear pose positive samples

(c) Negative image samples

Fig. 3 An example of the sample images from the image pool for vehicle detection

Table 1 The HOG parameter set configurations (fixed bin size = 8)

Config. No.	HOG parameters					Measured performance			
	WS	BS	CS	SS	Dimension	Time per Window (ms)	Accuracy rate (%)	Precision (%)	Recall (%)
1	32	16	8	8	288	0.33	93.5	99.6	92.2
2	40	16	8	8	512	0.45	96.5	99.9	99.8
3	40	24	8	8	648	0.61	96.0	99.8	95.2
4	48	24	8	8	1152	0.82	97.3	99.8	96.8
5	48	32	16	16	128	0.76	94.1	99.4	93.1
6	48	32	8	8	1152	0.95	95.8	99.9	94.8
7	48	16	8	8	800	0.62	96.9	99.9	96.2
8	**56**	**32**	**8**	**8**	**2048**	**1.37**	**97.8**	**99.9**	**97.4**
9	**56**	**24**	**8**	**8**	**1800**	**1.21**	**97.7**	**99.9**	**97.2**
10	56	16	8	8	1152	0.83	97.4	99.8	96.8
11	64	32	16	16	288	1.05	96.3	99.9	95.5
12	**64**	**32**	**8**	**8**	**3200**	**1.69**	**98.3**	**99.9**	**97.9**
13	**64**	**24**	**8**	**8**	**2592**	**1.33**	**97.8**	**99.9**	**97.3**
14	64	16	8	8	1568	0.99	97.2	99.8	96.7

In our experiments, the performance of classification is measured by metrics of accuracy rate, recall, and precision as shown in Eq. (2) through Eq. (4), where TP, TN, FP, and FN denote True Positives, True Negatives, False Positives, and False Negatives, respectively. We used five labels to categorize each vehicle pose and negative.

$$Accuracy\ rate = \frac{TP + TN}{TP + FP + TN + FN}. \tag{2}$$

$$Recall = \frac{TP}{TP + FN}. \tag{3}$$

$$Precision = \frac{TP}{TP + FP}. \tag{4}$$

The right side of Table 1 shows the experimental results with different parameters. The time per window means the sum of HOG extraction time and SVM classification time. The best accuracy rate is obtained as 98.3 % with parameter values of WS = 64, BS = 32, CS = SS = 8, BINS = 8 in Table 1. It is interesting that most of the precision are 99.8 or 99.9 % and thus, accuracy rate can be the representative metric for overall performance of classifier. The graph in Fig. 4 presents the overall relationship between dimension and accuracy rates or computing time. Note that except a few cases marked as thick arrows (config No. = 5 and 11 for time, No. = 3 and 6 for accuracy rate), the time per window is

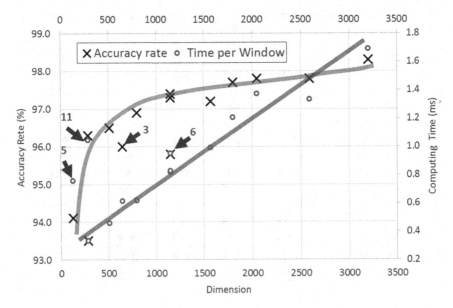

Fig. 4 The relationship between HOG dimension and accuracy rate or computing time

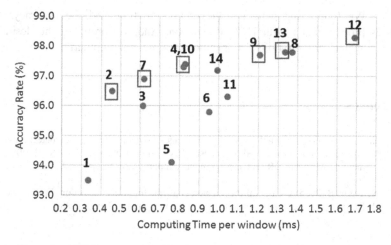

Fig. 5 The optimal HOG parameter set No. for each target precision (in *red squares*)

proportional linearly to the HOG dimension, and the accuracy rate is proportional logarithmically to the HOG dimension as shown in Fig. 4.

And, by inspecting Table 1 we can find the window size has the dominant contribution to the dimension and hence to the computing time and accuracy rate. This fact implies that we do not need to increment HOG dimension or window size beyond a certain number to get improved accuracy because higher dimension or bigger window size requires more computing cost than benefits from accuracy gain. It may be native to choose parameter set of config. No. = 12 for performance or, config. No. = 1 for computing time objectives. Instead, we should understand the problem-specific objectives and constraints in order to decide the best parameter value set.

Finally, we plotted the parameter sets in the space of accuracy and time per window in Fig. 5 to guide choosing the problem-specific best parameter set. In Fig. 5, for each target precision rate or time budget, we suggest the optimal parameter value sets marked as red squares with config No. of Table 1. For example, if the target minimum precision rate is 97 %, the config No. = 4 is the best choice. And, if the time constraints per window is 0.7 ms, the config No. = 7 is the best choice.

4   Conclusion

This paper aims to figure out desirable configuration of HOG descriptor for vehicle detection with SVM classifier. In this paper, we presented the relationship among HOG parameters and vehicle detection accuracy and average computing time per window. We found that generally the computing time and the accuracy rate increases proportionally with HOG dimension. And, we proposed the desirable HOG parameter configurations for specific constraints as summarized in Fig. 5.

References

1. Arrospide J, Salgado L, Camplani M (2013) Image-based on-road vehicle detection using cost-effective histograms of oriented gradients. J Vis Commun Image R 24:1182–1190
2. Mangasarian OL, Musicant DR (2001) Lagrangian support vector machines. J Mach Learn Res 1:161–177
3. Dalal N, Triggs B (2005) Histograms of oriented gradients for human detection. In: Proceedings of the IEEE conference computer vision and pattern recognition, June, pp 886–893

A Fast Algorithm to Build New Users Similarity List in Neighbourhood-Based Collaborative Filtering

Zhigang Lu and Hong Shen

Abstract Neighbourhood-based Collaborative Filtering (CF) has been applied in the industry for several decades because of its easy implementation and high recommendation accuracy. As the core of neighbourhood-based CF, the task of dynamically maintaining users' similarity list is challenged by cold-start problem and scalability problem. Recently, several methods are presented on addressing the two problems. However, these methods require mn steps to compute the similarity list against the kNN attack, where m and n are the number of items and users in the system respectively. Observing that the k new users from the kNN attack, with enough recommendation data, have the same rating list, we present a faster algorithm, TwinSearch, to avoid computing and sorting the similarity list for each new user repeatedly to save the time. The computational cost of our algorithm is $1/125$ of the existing methods. Both theoretical and experimental results show that the TwinSearch Algorithm achieves better running time than the traditional method.

Keywords Recommender systems · Neighbourhood-based collaborative filtering · Similarity computation · Database applications

Z. Lu · H. Shen (✉)
School of Computer Science, The University of Adelaide,
Adelaide, SA 5005, Australia
e-mail: hong.shen@adelaide.edu.au; hong@cs.adelaide.edu.au

Z. Lu
e-mail: zhigang.lu@adelaide.edu.au

H. Shen
School of Information Science and Technology,
Sun Yat-Sen University, Guangdong 510006, China

© Springer Science+Business Media Singapore 2016
J.J.(Jong Hyuk) Park et al. (eds.), *Advances in Parallel and Distributed Computing and Ubiquitous Services*, Lecture Notes in Electrical Engineering 368,
DOI 10.1007/978-981-10-0068-3_30

229

1   Introduction

In recommender systems, neighbourhood-based Collaborative Filtering (CF) is the most widely used method in the industry because of its easy implementation and high prediction accuracy [1].

The task of dynamically maintaining a similarity list is important in a neighbourhood-based recommender system, as the creation of new users and the rate updates of old users result in an update of the similarity list frequently. Accordingly, there are two main research problems in recommender systems, one is the cold-start problem, the other is the scalability problem.

Recent research [2–4] on addressing the cold-start problem focus on improving the prediction accuracy with the limit rates information. While, some of the solutions [2, 5–7] to the scalability problem work on decreasing the computational cost by linking the new similarity list with the old one.

Different from the two classic problems, we observe that the methods listed above do not work well against the kNN attack. In the kNN attack, the k new users (do not have similarity list yet) have enough rating data to build the reliable similarity lists, moreover, their ratings list are totally the same [8]. The methods aim to solve cold-start problem or scalability problem treat the kNN attack as a normal input of recommender systems. Thus, these methods require mn steps to compute each new user's similarity list, where m and n are the number of items and users respectively. Considering the number of users and items in a recommender system, the computational cost of the above methods is very large.

To address the problem of large computational cost caused by the kNN attack, we present a faster algorithm, TwinSearch, to avoid computing and sorting the similarity list for the new users repeatedly to save the time. Moreover, we compare the running time of TwinSearch algorithm and the traditional similarity computation method in two real-world data sets on both user-based and item-based CF. Both theoretical and experimental results show that the TwinSearch algorithm achieves better running time than the traditional method. To the best of our knowledge, we are the first to reduce the time complexity of building new users similarity list against the kNN attack in the recommender systems.

The rest of this paper is organised as follows: Sect. 2 presents our TwinSearch algorithm to build new users similarity list in neighbourhood-based CF recommender systems, and analyses the time complexity of TwinSearch theoretically. In Sect. 3, the experimental analyses of TwinSearch on the running time are provided. Finally, we conclude this paper with a summary in Sect. 4.

2 The TwinSearch Algorithm

2.1 Algorithm Design

In this section, we define the users who have the same rating list as *twin user*. To address the large computational cost against the kNN attack, we aim to avoid computing and sorting the similarity list for the new users repeatedly to save the computational resources. Since the new users are the same, our strategy is searching the twin user from the system, then copying the twin user's similarity list to the new user directly.

According to the properties of the similarity in recommender systems, if two users are twin user, i.e., $u_a = u_b$, we have the following relationships:

$$u_a = u_b \Rightarrow sim_{ai} = sim_{bi} \tag{1a}$$

$$u_a = u_b \Rightarrow r_{ai} = r_{bi} \tag{1b}$$

where sim_{ai} is the similarity between user u_a and u_i, r_{ai} is u_a's rating on item i. Relationship (1a) helps us to find the potential twin users from the system, Relationship (1b) helps us to find the exact twin user from the potential ones. Algorithm 1 shows how we compute the new user's sorted similarity list by finding and copying the twin user's similarity list.

Algorithm 1 TwinSearch Algorithm

Input:
　　A user-item rating set, R, with n users and m items; a user-user sorted similarity matrix, S
　　A new user, u_0, with several ratings on different items; a constant $c \in Z^*$.
Output:
The new user u_0's similarity list.
1: Select c random users, $u^*_i, i \in [1,c]$
2: **for** $i = 1$ to c **do**
3: 　compute similarity between user u_0 and u^*_i, sim_{0i};
4: 　search u^*_i's similarity list Si, for a $Set_i = \{u_x | sim_{ix} = sim_{0i}\}$;
5: 　**if** $sim_{0i} = 1$ **then**
6: 　　add u^*_i to Set_i;
7: 　**end if**
8: **end for**
9: Computer the intersection Set_0 of the c Set_is, $Set_0 = \cdot^c_{i=1} Set_i$
10: **for** $i = 1$ to $|Set_0|$ **do**
11: 　**if** $r_{ij} = r_{0j}, j \in [1,m]$ **then**
12: 　　copy the similarity list of $u_i \in Set_0$ to u_0;
13: 　　*break*;
14: 　**end if**
15: **end for**
16: **return** The new user u_n's similarity list.

In line 4, we search the potential twin users by Relationship (1a). In line 9, we narrow the size of the final potential twin user set Set_0 by intersecting the c bigger potential twin user set Set_i. The for loop in lines 10–15 find the twin user from the potential twin users' set by Relationship (1b). Our algorithm can be worked in both user-based and item-based CF, in this section, we present the TwinSearch algorithm from the perspective of the user-based methods, and this can be applied to item-based methods in a straightforward way.

2.2 Time Complexity Analysis

We select the c random users in line 1 in $O(c)$. The for loop in lines 2–8 contributes $O(c(m + \log n))$ to running time, if we use binary search in line 4. To compute the intersection Set_0 in line 9, it takes $O(cn)$ time. The for loop in lines 10–15 requires $O(|Set_0|m)$ time, if we use the link list as the similarity matrix S data structure. Therefore, the total running time of Algorithm 1 is $O(|Set_0|m + c(m + \log n))$.

Now we focus on the value of $|Set_0|$. Because $Set_0 = \cap_{i=1}^{c} Set_i$, $|Set_0| \leq \min\{|Set_i|\}$, i.e., $|Set_0| = \max\{\min\{|Set_i|\}\}$, $i \in [1, c]$. As the similarity values in a specific Set_i are equal, Set_i must be included in one sub-list of the original similarity list. The sub-list is produced by partitioning the similarity list with the similarity value. For example, suppose we have x sub-lists, then the similarity value in each sub-list is in the range of $[0, 1/x)$, $[1/x, 2/x)$, ..., $[1-1/x, 1.0]$ correspondingly. Thus, the upper bound of $|Set_0|$ must be less than the size of largest sub-list.

Moreover, Wei et al. [9] showed that any user's similarity list obeys a specific Gaussian distribution in recommender systems. Since for any Gaussian distributions, more than 99.99 % samples are in the range of $[\mu - k_4\sigma, \mu + k_4\sigma]$, we fix the similarity value range $[0, 1.0]$ within $\mu \pm 4\sigma$ in this paper. Figure 1 shows the basic statistic settings of one similarity list, where the greatest size sub-list's similarity value range is between $[\mu - k_3\sigma, \mu + k_4\sigma]$. Therefore, we have the size of the sub-list with the most number of users, s = (Area under the Gaussian distribution curve

Fig. 1 Distribution of user's similarity list

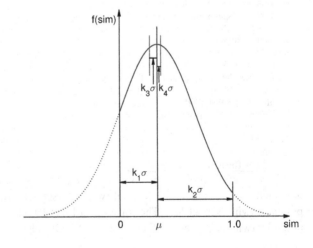

between $\mu - k_3\sigma$ and $\mu + k_4\sigma$/Area under the Gaussian distribution curve between 0 and 1.0) $\times n$.

According to the property of Gaussian distribution, we have:

$$\text{Maxmise}: s = \frac{\Phi(k3) + \Phi(k4) - 1}{\Phi(k3) + \Phi(k4) - 1} \times n \cdot$$

$$\text{Subject to } \mu - k_1\sigma = 0, \mu + k_2\sigma = 1$$

$$\mu - k_3\sigma = 0, \mu + k_4\sigma = 1/x \tag{2}$$

$$0 \le k_1 \le 4, 0 < k_2 \le 4$$

$$0 \le k_3, 0 < k_4$$

According to the properties of the cumulative distribution function of Gaussian distribution, the maximum solution to the linear programming (2) is $s = \frac{1}{125}n$, namely, $c \ll |Set_0| \le s = \frac{1}{125}n$. Therefore, the overall time complexity of TwinSearch is 1/125 of the traditional similarity computation method.

3 Experimental Evaluation

3.1 Datasets and Experimental Settings

In the experiments, we use two real-world datasets, MovieLens dataset and Douban film dataset. The MovieLens dataset consists of 100,000 ratings from 943 users on 1682 films. The Douban film dataset contains 16,830,839 ratings from 129,490 unique users on 58,541 unique films [10]. All the experiments are implemented in MATLAB 8.5 (64-bit) environment on a PC with Intel Core2 Quad Q8400 processor (2.67 GHz) with 8 GB DDR2 RAM.

3.2 Experimental Results

We design 4 experiments (Figs. 2, 3, 4 and 5) to evaluate the running time for k new user with same ratings on the above two data sets in both user-based and item-based CF. We use cosine similarity metric as the traditional similarity computation method, and set $k = 30$ in the 4 experiments. From the 4 figures, we can see that the TwinSearch algorithm achieves much better performance on time complexity than the traditional similarity computation method.

Fig. 2 Running time of
user-based CF on MovieLens

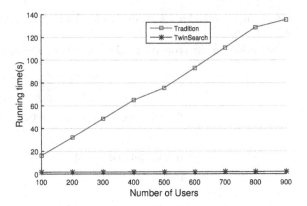

Fig. 3 Running time of
user-based CF on Douban
film

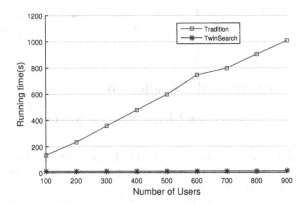

Fig. 4 Running time of
item-based CF on MovieLens

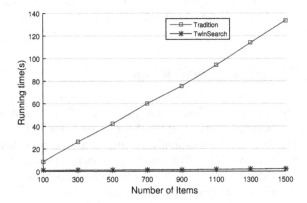

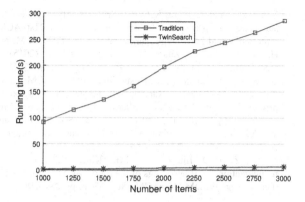

Fig. 5 Running time of item-based CF on Douban film

4　Conclusion

Neighbourhood-based CF plays an important role in e-commerce, because of the easy implementation and high recommendation accuracy. Two classic problems, cold-start problem and scalability problem, challenge the task of dynamically maintaining similarity list in neighbourhood-based CF. Recently, several methods are presented to solve the two problems. However, these methods applied an $O(mn)$ algorithm to compute the similarity list against the kNN attack. To address the problem of large computational cost, we present a faster algorithm, which avoids repeated computation to decrease the time complexity. Both theoretical and experimental results show that our algorithm achieves better running time than the traditional method.

Acknowledgements This work is supported by Australian Research Council Discovery Project DP150104871, Research Initiative Grant of Sun Yat-Sen University under Project 985, and National Science Foundation of China under its General Projects funding #61170232.

References

1. Liu B, Mobasher B, Nasraoui O (2011) Web data mining: exploring hyperlinks, contents, and usage data. Data-centric systems and applications. Springer, Berlin
2. Bobadilla J, Ortega F, Hernando A, Bernal J (2012) A collaborative filtering approach to mitigate the new user cold start problem. Knowl Based Syst 26:225–238
3. Liu JH, Zhou T, Zhang ZK, Yang Z, Liu C, Li WM (2014) Promoting cold-start items in recommender systems. PLoS One 9:e113457
4. Lika B, Kolomvatsos K, Hadjiefthymiades S (2014) Facing the cold start problem in recommender systems. Expert Syst Appl 41:2065–2073
5. Liu NN, Zhao M, Xiang E, Yang Q (2010) Online evolutionary collaborative filtering. In: Proceedings of the fourth ACM conference on recommender systems. RecSys'10. ACM, New York, pp 95–102

6. Yang X, Zhang Z, Wang K (2012) Scalable collaborative filtering using incremental update and local link prediction. In: Proceedings of the 21st ACM international conference on information and knowledge management. CIKM'12. ACM, New York, pp 2371–2374
7. Huang Y, Cui B, Zhang W, Jiang J, Xu Y (2015) Tencentrec: real-time stream recommendation in practice. In: Proceedings of the 2015 ACM SIGMOD international conference on management of data. SIGMOD'15. ACM, New York
8. Calandrino JA, Kilzer A, Narayanan A, Felten EW, Shmatikov V (2011) You might also like: privacy risks of collaborative filtering. In: Proceedings of the 2011 IEEE symposium on security and privacy. SP'11. IEEE Computer Society, Washington, DC, pp 231–246
9. Wei YZ, Moreau L, Jennings NR (2005) A market-based approach to recommender systems. ACM Trans Inf Syst 23:227–266
10. Ma H, Zhou D, Liu C, Lyu MR, King I (2011) Recommender systems with social regularization. In: Proceedings of the fourth ACM international conference on web search and data mining. WSDM '11. ACM, New York, pp 287–296

Printed in the United States
By Bookmasters